Creative Representations of Place

Cultural geography and the social sciences have seen a rise in the use of creative methods with which to understand and represent everyday life and place. Conversely, many artists are producing work that centres on ideas of place and space and utilising empirical research methods that have a resonance with geographers. This book contributes to the body of literature emerging from such creative approaches to place.

Drawing together theory and practice from cultural geography, anthropology and graphic design, this book proposes an interdisciplinary geo/graphic process for interrogating and re/presenting everyday life and place. A diverse set of research projects highlights participatory and autoethnographic approaches to the research. The sites of the projects are varied, encompassing the commercial space of grocery shops, cafés and restaurants, the private, domestic space of the home, and a Scottish World Heritage site. The theoretical context of each project highlights the transferability of the geo/graphic process, with place being variously framed within discussions of food, multi-culturalism and belonging; home, collecting and meaningful possessions; and materiality, memory and affect.

Themes in the book will appeal to researchers working in the creative methods field. This book will also be essential supplementary reading for postgraduate students studying Cultural Geography, Experimental Geographies, Visual Anthropology, Art and Design.

Alison Barnes is a senior lecturer at London College of Communication, University of the Arts London. Her practice-led research draws from both graphic design and cultural geography and centers on the understanding and representation of everyday life and place.

Routledge Research in Culture, Space and Identity

Series editor: Dr Jon Anderson, School of Planning and Geography, Cardiff University, UK

The *Routledge Research in Culture, Space and Identity Series* offers a forum for original and innovative research within cultural geography and connected fields. Titles within the series are empirically and theoretically informed and explore a range of dynamic and captivating topics. This series provides a forum for cutting edge research and new theoretical perspectives that reflect the wealth of research currently being undertaken. This series is aimed at upper-level undergraduates, research students and academics, appealing to geographers as well as the broader social sciences, arts and humanities.

For more information about this series, please visit: www.routledge.com/Routledge-Research-in-Culture-Space-and-Identity/book-series/CSI

Creative Representations of Place

Alison Barnes

Routledge
Taylor & Francis Group

LONDON AND NEW YORK

First published 2019
by Routledge
2 Park Square, Milton Park, Abingdon, Oxon OX14 4RN

and by Routledge
52 Vanderbilt Avenue, New York, NY 10017

First issued in paperback 2020

Routledge is an imprint of the Taylor & Francis Group, an informa business

2019 Alison Barnes

British Library Cataloguing-in-Publication Data
A catalogue record for this book is available from the British Library

Library of Congress Cataloging-in-Publication Data
A catalog record has been requested for this book

ISBN 13: 978-0-367-58878-6 (pbk)
ISBN 13: 978-1-138-06182-8 (hbk)

Typeset in Times New Roman
by Swales & Willis Ltd, Exeter, Devon, UK

For my Dad, who would have been quietly chuffed.

Contents

Figures

Acknowledgements

Many people have contributed to the emergence of the ideas that have led to the development of this book and have supported my journey along the way. My thanks must first go to Russell Bestley and the late Ian Noble who provided a stimulating space for testing and challenging the boundaries of graphic design and typography across both theory and practice. It was with their encouragement that the foundations for this research took shape. During my PhD, the help and advice of my supervisory team, Teal Triggs, Phil Baines and Catherine Nash, was invaluable, as was the support of my University of the Arts 'space and place' colleagues Pat Naldi, Debbie True and Fay Hoolahan. My thanks also go to the many participants who have engaged with my research and offered me insights into their lives in Hackney. The financial support to undertake the PhD was provided by a three year Arts and Humanities Research Council doctoral award, and I am hugely grateful to have been given this opportunity. Post PhD, many friends and colleagues have been pivotal in helping these ideas to further develop and in particular I'd like to thank Ben Coles, Michelle Catanzaro and Samantha Edwards-Vandenhoek for productive conversations and collaborations. I am also grateful to colleagues at London College of Communication for supporting my sabbatical which enabled the writing of this volume, and to Graham Goldwater in particular for his help with the photography. I would also like to thank my assistant editor at Routledge, Ruth Anderson, and Barbara Spender, the most eagle-eyed of copy editors. Finally, I must also thank my partner, Cath, for her patience and unwavering belief in me, and the kittens whose contributions via a quick walk across the keyboard were interesting typographically, but rarely on point.

Introduction

The past several years have seen a rise in the use of creative methods with which to understand and represent everyday life and place within cultural geography and the social sciences more broadly. Conversely, many artists are producing work that centres on ideas of place and space and utilising empirical research methods that have a resonance with geographers. This book contributes to the growing body of literature emerging from such creative approaches to place. Adopting a 'geo/graphic' approach that draws together theory and practice from cultural geography, anthropology and graphic design, it proposes an interdisciplinary methodological approach for interrogating and re/presenting everyday life and place that is underpinned by a similarly interdisciplinary theoretical framework. The potential of a geo/graphic approach to engage with place in a variety of different ways and settings is highlighted by the diversity of the projects discussed in Chapters 5, 6 and 7 of the book. These include both participatory and autoethnographic approaches to the research; projects that have been undertaken within the private, domestic space of the home; and those that focus on the public space of the street or the commercial space of grocery shops, cafés and restaurants. The theoretical context of each project highlights the transferability of the geo/graphic process, with place being variously framed within discussions of food, multi-culturalism and belonging; home, collecting and meaningful possessions; and materiality, memory and affect.

The projects and approach discussed within this book build on developments within geography, anthropology and graphic design over several decades. For example, the creative and spatial 'turns' evident across a range of disciplines that have led to this rise in creative methods go hand in hand with a further series of interlinked turns; a practice turn, a bodily turn and a reflexive turn. A series of theoretical developments have contributed to this particular surge in interest in both place and creative methods. Firstly, in the late 1980s, postmodernism and the 'crisis of representation' led to the shattering of the supposed one-to-one relationship between writing and reality. This inevitably focused attention onto the perspective of the writer and in doing so, prompted the development of a reflexive approach. Whilst much work was done to counter hegemonic positions within both geographic and ethnographic work, postmodern thinking, and latterly poststructuralist, still positioned the idea of a representational text as problematic. This, and contemporary conceptions of place which position it as ongoing, relational and always in flux, led to the development of non-representational thinking. Drawing, in part, from earlier humanistic geography and associated phenomenological approaches non-representational work brought an increased focused on bodily experiences and practice in the context of place.

There are now many spaces for dissemination of such work and many opportunities for collaboration, particularly of an interdisciplinary nature – most predominantly that of art-geography. However, the use of creative methods is an emerging field, and many writers contend that there is still much to be done to both share practice and in turn further develop such approaches. Rather than a set of pre-packaged methods or a 'how to' manual, this book offers those from disciplines such as geography, anthropology, art and design a point of entry through which they can further develop their own approaches and disciplinary perspectives that in turn will further this exciting field.

Interdisciplinarity, art-geography and creative methods

Such creative approaches and collaborations, by their very nature, require and encourage inter-, cross- and multi-disciplinarity. The ability to synthesise knowledge and methods from different disciplines, or at least be in a position to view one discipline from the perspective of another, can be very productive, enabling one to reframe ideas and methods in the context of unfamiliar theories and practice. Cocker (2008: 1) describes an interdisciplinary approach as one that signals 'a position of inbetweenness – of both being within and without – of maybe playing the game but with different rules'. This idea of 'without' has been further described as undertaking a project that 'isn't organised around an . . . existent methodology' (Phelan & Rogoff 2001: 34). This should not be seen as 'a form of lack' or 'turning your back on, or denying what you had at your disposal previously'. Rather it assumes that one had a model to begin with that is no longer relevant or useful, so 'one is actively doing without the certitudes', without as yet 'having produced a hard-and-fast subject or methodology to replace them' (Phelan & Rogoff 2001: 34). However, for some, this position of 'inbetweenness' or 'without' can produce a paralysing uncertainty or a methodological and disciplinary defensiveness (Nash 2013: 52), so it is not without its challenges.

To date, within the context of geography, these interdisciplinary collaborations, the use of creative research methods, and representational strategies that seek forms other than the academic journal, have largely drawn from art and art practice. Geography has a history as a visual discipline, and perhaps the image that first springs to mind in relation to geography is that of a map. However, geographers have been at the forefront of developing 'visual methodologies' (Rose 2016) and analysing creative forms such as landscape painting, films and creative writing in terms of how they represent place. This current body of creative research builds on those foundations. It also stems from contemporary definitions of place. Place was seen by humanistic geographers as having a uniqueness, a specific 'genius loci' or sense of place that foregrounded 'localness' and engendered metanarratives that excluded those who were seen as outsiders. Geographers such as Doreen Massey (1994, 2005) repositioned place as relational, as a product of both local and global flows that come together to form places that are both interconnected and unique. Seen in this context, place is therefore always in process, constantly being formed and reformed, and is always, therefore, unfinished (Massey 2005: 107). This view clearly creates an issue for the researcher seeking to engage with place. To understand and represent place, one inevitably steps into the flow of events and in some respects 'freezes' these at a moment in time in order to analyse and theorise. Such a representational stasis effectively goes against the grain of relational perspectives on place. To this end, non-representational theory has urged geographers, amongst others, to engage with methods more creatively in order to reflect this flow rather than restrict it. Many of these approaches have therefore utilised media and methods that are perceived as offering both a process and practice that are naturally more 'open' and therefore capable of contending with such issues. For example, contemporary cultural geographers have worked with practices such as dance, performance, curation, film, sound, and the spoken word, all of which release geography and geographers from the tyranny of written language and the perceived fixed nature of the printed word. These types of methods, and many others, are usually discussed and framed within the context of arts practice in the widest sense. Much has been written about art's capacity for developing work that is performative (Crouch 2010) and relational (Bourriaud 2002) along with its potential to provide geography and geographers with creative approaches for understanding and disseminating everyday life and place (see for example Hawkins 2015; 2013; 2011).

Much of this commentary is one way, focusing on art's contribution to geography, rather than geography's potential contribution to such interdisciplinary collaborations (Jellis 2015: 369). There is also rarely any mention of the potential of design to contribute to the development of such creative approaches, which, given the increasing uptake of a 'design thinking' approach within areas such as health and business, seems an oversight, though this may be partly related both to perceptions of design as a service industry and its relatively youthful academic profile. For example, degrees such as graphic design were only offered in the UK's so-called 'new universities', many of which are former polytechnics that were granted university status, from 1992. Outside of higher education the phrase graphic design wasn't coined until 1921, so in contrast to subjects like fine art, it has also had much less time to develop in terms of what it might offer beyond the commercial realm. However, Tolia-Kelly (2012: 139) suggests that geographers might benefit from reframing such creative work within the context of 'visual culture . . . rather than a self-consciously lived, artistic impression'. In doing so she suggests that the practices inherent in visual culture 'are more about pattern, design and impression' and would provide a more appropriate location for 'geography's visual research edge' (Tolia-Kelly 2012: 139). A geo/graphic approach is therefore one that positions itself at this edge, and in doing so offers new strategies for those interested in creative methods that engage the reader in re/presentations of place that are able to capture its relational and complex nature and utilise written text and the printed page.

A geo/graphic approach

Before going any further I should perhaps clarify the context in which I have coined and am using the term geo/graphic. I do not see it as a neologism, for I am not intending it to offer a new sense of an existing word. The word geography can be literally translated as earth writings, emerging from its Latin roots of geo – meaning earth, and graphy – meaning writing. By taking the word geographic and inserting a forward slash between the two parts of the word, its constituent parts – geo and graphic – are re-emphasised. Graphic these days is perhaps associated more with pictures than with words and relates to graphic design, an inherent part of the geo/graphic approach. This shifts the interplay of meaning to a more visual perspective and points towards the interdisciplinary nature of the work. The use of a forward slash as opposed to a hyphen was a particular choice. It has been described as a character that can perform a simultaneous 'division and doubling' and 'hint at meaning that is not quite there yet' (Springgay, Irwin & Wilson Kind 2005: 904). However, the slash doesn't allow limitless interpretations in relation to these new meanings.

> It is the tension provoked by this doubling, between limit/less that maintains meaning's possibility . . . The slash suggests movements or shifts between the terms . . . The slash makes the terms active, relational, as they reverberate with, in, and through each other.
>
> (Springgay, Irwin & Wilson Kind 2005: 904)

The angle of the forward slash allows for this movement, each word moves both towards and away from the other, creating an interactive coupling with the slash acting as a fulcrum or pivot (see Barnes 2012a). In the context of geo/graphic, the slash highlights this between space as a productive new place to be found between the disciplines of geography and graphic design. The space created by the slash could also be seen as a kind of fold, with graphics and geography folding into and around each other: 'Folding holds out the potential to diversify endlessly without falling into the logic of binary oppositions. This sense of the fold thinks of matter as doubling back upon itself to make endless new points of connection between diverse elements' (Meskimmon 2003: 167). To that end, the forward slash is used to reconfigure the context of the word representation in discussion of creative outputs that endeavour to go beyond a one-to-one 'mapping' of place. The use of re/presentation in relation to both the research and practice of this type emphasises both 're' and 'presentation' and again creates a productive interplay that enables one to move beyond the idea of the mimetic with regard to an image of place. The separation of 're' enables its root meaning of 'anew' or 'afresh' to be foregrounded, but it also opens up the sense that such work is 'about' ideas and possibilities of presentation itself.

In developing a framework for a geo/graphic approach, the primary disciplines drawn from are geography, anthropology and graphic design. Geography in terms of theoretical approaches to place and representation; anthropology in terms of an ethnographic approach to the understanding of everyday life and place; and graphic design in terms of design as a method of inquiry and the potential inherent in graphic design and typography to develop print-based re/presentations of place that no longer 'silence the page' (Ingold 2007: 24–26) or position the reader above and outside of the territory. In drawing these facets together, geo/graphic work engages with 'writing' in two ways. In researching place, a geo/graphic approach brings together a range of 'texts'; for example, field notes, field writing, documentary photographs, participants' photography, conversations, interviews, documentary and archival research materials, and maps. These texts cumulatively build narratives of and within place and are then analysed, assembled and re/presented as a 'geo/graphy'. In the process of developing this re/presentation, graphic design is actively used both as a mode of inquiry and to develop a proactive space of communication and interpretation for the reader.

Unlike art, which is often framed in the context of self-expression and individual genius, graphic design is underpinned by the 'design process' – a flexible, iterative and recursive process that could be likened to a process of analysis. Throughout the design process, a designer's decisions can be clearly articulated, and in a sense 'validated'; therefore the process offers a useful accessibility. It also offers transferability, as the process of graphic design is generic in nature, but cannot be undertaken until one has content and context – these could be anything at all, therefore the process is inherently adaptable (see Barnes 2012a). Graphic designers are also used to bridging the divide between written content and visual representation and do this in a way that endeavours to communicate to a specific audience. Thus, they work with context and content in order to develop a concept for communication. All facets of these aspects of the project come to bear on the final form, and graphic designers make proactive choices as to tactile properties such as materials, form and binding; visual properties such as colour, typefaces and imagery; textual properties such as typographic hierarchy, copywriting and typographic expression; and even olfactory properties such as smells which can be added to paper stock. In the terms of print based work, it is therefore entirely possible to develop a multi-sensory experience for the reader. Through the utilisation of non-traditional layout with text and imagery, changes in pace through multiple pages, and the inclusion of hidden elements within folds, or items that require the reader to handle or reconfigure the design in some way, graphic designers also have the ability to create print-based work that is interactive and non-linear. This combination of possibilities – multi-sensory experience, non-linearity, and interactivity – therefore posits print-based work as having the potential to engage with relational conceptions of place and ideas of the non-representational. A geo/graphic approach therefore offers geographers – and others interested in the use of creative methods from outside the disciplines of art and design – an opportunity to reconsider the potential of text and the printed page. It also offers an insight into the role design and designers might play going forward in such methodological and collaborative developments. Furthermore, it offers designers and artists an opportunity to engage with theoretical and methodological discussions from both geography and anthropology. This will strengthen their understandings of place and representation, and provide them with tools to position their work more formally within the context of an ethnographic methodology. Whilst this is an approach designers and artists use on a regular basis, they often do so without reference to ethnography specifically, as undergraduate art and design education rarely engages with such methodological frameworks (Barnes 2012b).

The structure of the book

The first four chapters of the book frame the theoretical and methodological concerns that underpin the geo/graphic process and the final three chapters discuss geo/graphic projects that show the transferability and adaptability of the methodology. Chapter 1 opens the book with a discussion of place and everyday life. Contemporary relational definitions of place are framed within a trajectory of thought stemming from humanistic geographers of the 1970s. Place is a word often used in a simplistic, unquestioning way in everyday conversation, yet it is revealed as a complex term, one that positions place as both interconnected and unique. Both place and everyday life are co-constitutive of each other, and everyday life, much like place is continually emergent. Chapter 2 focuses on the representation of place, with the iconic geographic image of the map offering a starting point for discussion. However, maps, along with all other forms of supposedly scientific and objective representation were challenged in the late 1980s during the 'crisis of representation' – language was revealed as inherently unstable and the position and perspective of the researcher were implicated in the rhetorical construction of the text. At the turn of the twenty-first century, non-representational theory was posited as a way of contending with many of these issues and with contemporary relational conceptions of place. Many of those undertaking non-representational work sought new ways to engage with place and eschewed methodological conservatism, thus Chapter 3 opens with a discussion of such creative methods and various art-geography collaborations. It continues by introducing the discipline of graphic design and highlighting the opportunity to move beyond superficial discussions of style to a more ambitious agenda that has the potential to challenge Ingold's (2007) perception that the page has been silenced by the advent of mechanical print. Chapter 4 brings these various theoretical strands together in outlining a geo/graphic methodology that draws together traditional ethnographic approaches with visual and sensory ethnographic work and the practice of walking. It frames writing and design as simultaneously modes of inquiry and modes of telling, and offers a simple framework for analysis that reflects and complements an approach underpinned by the design process. Taken together, the first four chapters of the book position a geo/graphic approach as a creative intervention and re/presentation, both into and of place, that is driven by a qualitative, naturalistic and reflective methodology using multiple methods with which to gather and analyse empirical research.

The final three chapters discuss geo/graphic projects that have each been undertaken in a specific context using a particular set of methods to form the geo/graphy. Chapters 5 and 6 both centre on the London borough of Hackney and explore the potential of the book to re/present place. Chapter 5 focuses on food, consumption and multi-culturalism, and in doing so highlights the book as capable of re/presenting an interactive, non-linear, spatial experience of place that is both local and global. Chapter 6 explores what it is that makes a house a home and, through the use of participants' stories and memories, alongside the use of printed ephemera and non-traditional forms of construction, highlights the book as having the potential to re/present an emotional, personal, experiential journey through place. Chapter 7 features a geo/graphic project undertaken in Edinburgh Old Town that, through the use of scale, colour, historical texts and reader interaction, positions the book as an embodied, affective experience of place. Each of these three chapters culminates in a discussion of the project's context, emphasising that the geo/graphic process is not solely concerned with re/presentation, but with the understanding of place also. Drawing the book to a close, the conclusion reiterates the potential of a geo/graphic approach across multiple disciplines; the potential of interdisciplinary perspectives and multi-disciplinary collaboration; and looks forward, highlighting areas for further development. In particular, it highlights the need for the development of projects that use such creative methods for more critical ends, to challenge the perception that much creative work is apolitical at the very least and, at its worst, encouraging a return to a white male colonial perspective (Hawkins & de Leeuw 2017). However, first we turn to place, and the development of a conception and definition of place within contemporary geography that underpins both the theory and practice throughout the rest of the volume.

Bibliography

Barnes, A. (2012a) 'Thinking geo/graphically: The interdisciplinary space between graphic design and cultural geography', *Polymath: An Interdisciplinary Arts and Sciences Journal*, 2(3), pp. 69–84.

Barnes, A. (2012b) 'Repositioning the graphic designer as researcher', *Iridescent*. 2(1), pp. 3–17.

Bourriaud, N. (2002) *Relational Aesthetics*. Dijon: Les Presses du Réel.

Cocker, E. (2008) 'Wandering: Straying from the disciplinary path' *Creative Interdisciplinarity in Art and Design Research: Interrogations Workshop*, De Montfort University, Leicester, 21 January. Available at: http://wandering-straying.blogspot.co.uk/ (Accessed: 29 December 2017).

Crouch, D. (2010) 'Flirting with space: Thinking landscape relationally', *cultural geographies*. 17(1), pp. 5–18.

Hawkins, H. (2015) 'Creative Geographic Methods: Knowing, Representing, Intervening', *cultural geographies*. 22(2), pp. 247–268.

Hawkins, H. (2013) *For Creative Geographies: Geography, Visual Arts and the Making of Worlds*. Abingdon: Routledge.

Hawkins, H. (2011) 'Dialogues and Doings: Sketching the Relationships between Geography and Art', *Geography Compass*. 5(7), pp. 464–478.

Hawkins, H. & de Leeuw, S. (2017) 'Critical geographies and geography's creative re/turn: Poetics and practices for new disciplinary spaces', *Gender, Place and Culture*. 24(3), pp. 303–324.

Ingold, T. (2007) *Lines: A Brief History*. Abingdon: Routledge.

Jellis, T. (2015) Spatial experiments: Art, geography, pedagogy', *cultural geographies*. 22(2), pp. 369–374.

Massey, D. (2005) *For Space*, London: Sage.

Massey, D. (1994) *Space, Place and Gender*, Minneapolis: University of Minnesota Press.

Meskimmon, M. (2003) *Women Making Art*. London: Routledge.

Nash, C. (2013) 'Cultural geography in practice' in Johnson, N., Schein, R. H. & Winders, J. (eds) *The Wiley-Blackwell Companion to Cultural Geography*. Chichester: John Wiley & Sons, pp. 45–56.

Phelan, P. & Rogoff, I. (2001) '"Without": A conversation', *Art Journal*. 60(3), pp. 34–41.

Rose, G. (2016) *Visual Methodologies*. 1st edn 2001. London: Sage

Springgay, S., Irwin, R. I. & Wilson Kind, S. (2005) 'A/r/tography as living inquiry through art and text', *Qualitative Inquiry*. 11(6), pp. 897–912.

Tolia-Kelly, D. (2012) 'The geographies of cultural geography II: Visual culture', *Progress in Human Geography*. 36(1), pp. 135–142.

1 Defining place

Introduction

Whilst place is clearly central to the discipline of geography, it has become increasingly important throughout the social sciences and humanities in what has been described as a 'spatial turn' (Hubbard & Kitchin 2011: 2). Although much of this chapter frames the definition of place within geographical thought, place is not solely the province of geographers, it underpins a range of disciplines, and, in terms of its study, 'benefits from an interdisciplinary approach' (Cresswell 2015: 1). In writing about the practices of everyday life, anthropologist and ethnographer Sarah Pink expresses similar thoughts about interdisciplinarity and draws from philosophy, anthropology, sociology and geography in framing her approach (Pink 2012: 1–2). Similarly, this chapter, and the book as a whole, draw from such disciplines in framing the theoretical and methodological discussions that underpin both place and everyday life.

The word 'place' is used regularly in everyday language – your place, my place, that place – and because of this common usage its meaning is rarely questioned. However, place – in geographic terms – is a complex proposition and the first section of this chapter outlines the development of a definition of place that is used to underpin the book. Place is not simply thought of in terms of a location or coordinates on a map, places are sites of meaning, and therefore are 'a way of understanding the world' (Cresswell 2015: 18). Unlike space, which remains unpeopled and framed within ideas of volume and the quantifiable, place is peopled, and more often than not discussed in qualitative terms. Humanistic geographic thinking in the 1970s drove this shift from a scientific, rational approach to the spatial, to a repositioning of people as central to geography. Geographers such as Yi Fu Tuan focused on the particularity of place as much geographic work moved from a nomothetic approach to an ideographic. This focus on people, place and experience foregrounded ideas of belonging and place-attachment.

However, Marxist geographers such as David Harvey, also working in the 1970s, began to frame places as interconnected and interdependent. Harvey proposed that, largely driven by the emergence of a global economy, places were becoming less and less unique, to the extent that ideas of 'placelessness' and 'nonplaces' were posited by geographer Edward Relph and anthropologist Marc Augé respectively. Relational geographers such as Doreen Massey developed a further, more nuanced approach to place in the context of globalisation that enabled a framing of place as both interconnected and unique. Thus, global flows and networks converge in places in different and specific ways, creating a global sense of the local. This 'uniqueness' of place as posited by both humanistic and relational geographers could be said to be formed by the products and practices of everyday life – these emerged as an area of intellectual inquiry after the Second World War, and are currently resurgent as an area of focus within geography. Both place and everyday life are inextricably linked, reflecting, refracting, producing and reproducing each other. It is through our everyday actions and practices that place is constituted; it is the very 'dailyness' of life in urban contexts – 'routines, habits, behaviours and objects' – that enables 'much of city life to cohere' (Latham & McCormack 2007: 25). Because of its very nature – its often mundane, familiar, routine practices – everyday life might be in danger of being overlooked or seeming irrelevant. Therefore, many approaches to the study of everyday life attempt to make the familiar strange or the ordinary extraordinary – the approach of the avant-garde Surrealist movement being a prime example of this.

Defining place

In general conversation the terms 'place' and 'space' are often used interchangeably, with little thought as to the specific differences between them, or the subtler meanings of either. Dictionary definitions of space define it in terms of a continuous area or expanse that is available or unoccupied; in relation to quantifiable or measurable terms relating to volume or time; or in the context of positioning things apart or away from each other. Space, therefore, seems to be about absence, and is often discussed within a rational, scientific framework that seeks to define things quantitatively. The primary dictionary definition of place is that of a specific position within space, or the act of putting something in a particular spot. In other words, place is primarily defined as being about location and in this context is seemingly a more specific version, or subset, of space. However, the secondary dictionary definition of place refers to it as a portion of space that is available to be used by someone; for example, a seat on the bus, a place at a table, or a desk in an office or classroom. This 'peopling' of place and the association of ownership, or perhaps belonging, brings us closer to understandings of place as discussed and utilised within this book.

In human geographic terms, the term place belies one of the most complex ideas within the discipline, with contemporary understandings of place positioning it as a meaningful location, defined in three fundamental, interlinked ways with location one constituent part of this triumvirate. For Agnew (2005), place is also discussed in relation to locale, which is essentially the setting and scale for our everyday actions and interactions. Locale differs from location in that it is not about position, rather it is a focus on the material, visual form of a place. The third aspect of this definition is that place is also framed within the idea of a sense of place and in this understanding every place is particular and therefore singular. The concept of a sense of place relates to a sense of belonging and participation

(Agnew 2005: 89). It therefore relates to the way people feel about places and how places play a role in the construction of personal identity and that of particular groups related to a specific place in some way. A similar proposition is offered by Gieryn (2000), who suggests that location is complemented by both material form and meaningfulness. He also states that these three elements cannot be judged as one more important than the other and nor can they be separated. The 'completeness' of place is destroyed if one were to be discarded (Gieryn 2000: 466).

It is immediately apparent from these two more fleshed out conceptions of place that people and place are inextricably linked. Unlike space, which can be literally defined as unoccupied, place is populated and both those occupying place – and place itself – are reciprocally constructed. Rather than being a subset of space, place is therefore 'the counterpoint of space' (Anderson 2010: 41). When moving somewhere new, whether that be a house or flat, or a desk within a workplace, we personalise our territory, we add our own possessions, we decorate to our own tastes – we transform an address or spatial location into our own place. When we think of a place important to us often the first things that come to mind are our childhood home or the place of our birth – places that are usually linked with a sense of belonging and meaningfulness, places that play a part in defining our identity. More often than not, we develop close emotional ties to these places, regardless of their different scales. Incrementally, they become imbued with meaning which often lasts long beyond the actual time we may spend living somewhere. We don't think of the dimensions of these spaces (in fact we often misremember these, imagining childhood spaces as bigger than they actually were), rather we often remember these meaningful places through the sound of a squeaky cupboard door, the smell of our favourite meal being cooked, or the touch of a particular surface. The framing of a house or particular childhood places in this way could be described as nostalgic or romantic, and as we will see ideas of home have often been conflated with ideas of safety and security even though reality can often be different. However, this type of experiential reading of place has its roots in theoretical work developed by humanistic geographers several decades ago. In turn, their approach was rooted in the work of philosophers such as Martin Heidegger and Maurice Merleau-Ponty, and place was seen as a way of being in the world, as 'a universal and transhistorical part of the human condition' (Cresswell 2015: 35)

Humanistic geographers' approaches to place

Modern geographical understandings of place have been debated and developed since the 1970s and were prompted by the emergence of humanistic geographers who rejected the postwar scientific and quantitative approach (Castree 2003: 157) and the emphasis placed on space (Cresswell 2015: 35). This shift therefore positioned humans at the centre of geographic thought and privileged the use of qualitative methods with which to understand people, place and everyday life. Humanistic geographers emphasised the particularity of place experience rather than seeking overarching similarities and patterns. For example, Yi Fu Tuan describes this specificity of place as 'made up of experiences, mostly fleeting', being a 'unique blend of sights, sounds and smells' (Tuan 2007: 183) and likens place to pauses (Tuan 2007: 6). The pauses enable us to know place better and inscribe our own meanings and value onto it, to develop a 'feel' of a place 'in one's muscles and bones' (Tuan 2007: 198). For humanistic geographers, place wasn't just about location, place had meaning (Cresswell 2013: 112) and geographers such as Tuan replaced the law-finding nomothetic approach with the ideographic search for the uniqueness of place, or its 'genius loci' (Crang 1998: 101).

Through this humanistic approach, an inherent geographic interest in places was further developed by an emerging concept of place that foregrounded the subjective experience of people. This change in thinking radically shifted human geography from a previous focus on spatial science, objectivity, and the idea of people as rational beings, to a field of study that focused on 'the relationship between people and the world through the realm of experience' (Cresswell 2013: 113). Tuan's belief is that we understand the world though our experience of places and similarly Relph states that 'to be human is to live in a world that is filled with significant places: To be human is to have and to know *your* place' (Relph 1976: 1; italics in original). This quote clearly articulates the centrality of place-making and place attachment to the human condition (Price 2013: 121). In a sense this is about putting down roots and feeling secure, thus the concept of home is often thought of in connection with such feelings.

Geographers such as Tuan and philosophers such as Heidegger and Bachelard have helped establish the concept of home as central to humanistic approaches to place. Home is seen as a safe, secure, intimate and personal space, private as opposed to public, and offering control and freedom over what happens within it. For Bachelard, home is the first place we understand and we therefore frame our consequent understandings of the external world within this context. Bachelard also suggests our memories are made and fixed within place, rather than in time as more commonly thought, and it is the house that draws out the 'topography of our intimate being' (Bachelard 1994: xxxii). It should be pointed out that much work has since been done to counteract what could, in isolation, be perceived as an idealised, romantic view of place that for some people doesn't exist. Places can exclude, be precarious and be unsafe – for example, contemporary feminist geographies focusing on the space of the home for women reveal it to be far from a site of safety and security for those experiencing domestic violence. Rose (1993) argues that, for many women, home can be a site of conflict rather than care, of oppression rather than freedom. Rose therefore also calls into question the inextricable link between ideas of belonging and home and the overarching sense that home is synonymous with the concept of place.

Similarly, the humanistic geographers' conception that it is possible to discover a specific and authentic genius loci of place has also since been dismissed. Like home, place is inevitably experienced by different people in different ways. The idea of some kind of hidden truth being embedded within place is now seen as nostalgic, as a romanticising or cleansing of the past. By adopting this approach, it effects a 'denial of difference', one that excludes any kind of particular or individual experience of place and assumes a 'universal subjectivity' – that we all share the same genius loci inherent within place. This inevitably leads to the voice of marginalised groups being overlooked or forgotten, and the stories or 'truths' that become embedded within place are those of more powerful social groups (Holloway & Hubbard 2001: 112). In effect they establish a dominant metanarrative which overrides the narratives of those less able to be heard. One criticism of humanistic geography, therefore, is that it does not critically engage with the inequalities evident within many people's experiences of place that are the result of 'power-laden differences' in relation to class, sexuality, gender or ethnicity, for example (Holloway & Hubbard 2001: 113).

This critique of the romanticised concept of home and the issues of a genius loci notwithstanding, many of the further theoretical approaches and methods discussed in the following chapters owe a debt to the humanistic geographers of the 1970s and many of their ideas about place have been subsumed within the contemporary discipline of geography (Creswell 2013: 119). However, during the 1970s, the development of a global economy led Marxist geographers to argue that places were both interconnected and interdependent, and therefore these global connections between places were deemed more important than differences (Castree 2003: 158–159). Globalisation also began to be seen as a force that homogenised place. Going further still, Augé's (1995) proposition of 'non-places' – a direct product of globalisation – suggests that traditional understandings of place are becoming less important and are being replaced by places that are transient. These 'spaces of circulation' include freeways and airways as well as places like department stores and supermarkets – 'spaces where people coexist or cohabit without living together' (Augé 1995: 110). Such places also call to mind Relph's (1976) discussion of 'placelessness', an increasing 'inauthenticity' evident in these types of ubiquitous places and our relationships with them. If we have travelled through a major international airport recently, it is very likely that we can understand both these positions. For example, Dubai airport – the world's busiest international passenger airport – presents a strangely disorientating experience for the traveller. This is an experience where the idea of place is clearly recognised as important but the sense of place conjured up by the variety and choice of cafés and restaurants reflects the global realm of travellers' home territories, rather than one that is rooted within the indigenous social and cultural context of the United Arab Emirates. The familiar material quality of brands like Le Pain Quotidien or Camden Food Co. with their particular colours, specific shop fittings and furniture allows us to experience place in Dubai as we might at home. A sense of place in Dubai Airport is chameleon like – it becomes what we each need it to be as we pass through it, whether we are seeking the familiarity of Paris, London or beyond. One might argue that Dubai, as with all places, is socially constructed and experienced subjectively by all who pass through. However, the majority of places evolve over time and develop in less planned ways than an international airport terminal. This deliberate attempt to build a globally recognisable destination leads to the development of a place that seems removed from its geographic location, thus isolating one of the tripartite components of the definition of place.

Rather than space being transformed into place here, this is perhaps an example of Harvey's suggestion that capitalism actually reverses that process, transforming place back into space (Harvey 1996) or at least facilitating a view of place that positions it as more like space. Therefore, late capitalist visions of place define it in terms such as 'abstract, smooth, masculine, cerebral and unanchored' (Price 2013: 121) – a world away from the humanist perspective. However, Massey (1994, 2005) proposed a more nuanced view of place that, whilst accepting the presence of global forces, also enables a sense of the local.

Relational geographies of place and space

Whereas a humanistic geographic approach focused on the specific essences of places and the world constructed through a series of discrete entities, a relational approach understands the world as formed through the way in which things relate to each other. In some senses this can be seen as a topological rather than topographical approach as it is one where scale and location are less important than points of connection (Cresswell 2013: 218). In this perspective, the world is framed in terms of networks and flows that link 'any one local place to a host of other places the world over' (Crang 1999: 31). Whilst earlier views of globalisation saw it as eroding place and locality, Massey's conception of place connects the global and the local, offering 'a global sense of the local' (Massey 1994: 51) that allows us to view places as both interconnected and unique. It is the particular way that such local and global networks and flows interact and intersect in a place that results in its uniqueness. In contrast to Tuan's notion of 'pause', Massey (1994, 2005) defines place as 'process', as something that is open, not static. For Massey, place is 'the sphere of a dynamic simultaneity, constantly disconnected by new arrivals, constantly waiting to be determined (and therefore always undetermined) by the construction of new relations. It is always being made and always, therefore, in a sense unfinished' (Massey 2005: 107). Massey's (1994: 152–153) description of the London neighborhood of Kilburn paints a vivid picture that is simultaneously ordinary and extraordinary, local and global – a heterogeneous, multi-cultural, richly chaotic tapestry of life that is likely to be recognisable to anyone living in a large city today. For example, the London borough of Hackney, where two of the creative re/presentations of place featured in later chapters were developed has evident similarities. In Hackney, food shops and restaurants cater to a diverse ethnic population; sought after enclaves exist where Victorian houses cost upwards of three million pounds; run down, sprawling housing estates sit alongside gentrified Georgian terraces; and loft-style living apartments have sprung up on the back of the 2012 Olympics and Dalston's new found fame as one of the capital's cultural hotspots. These elements and other interpretations of Hackney are meaningful in different ways to different members of the community.

> Hackney is distinct—for the old white working class, for the variety of ethnic minorities, for the new monied gentrifiers. Each has its own view of what the essential place is, each partly based on the past, each drawing out a different

potential future . . . Hackney *is* Hackney only because of the coexistence of all of those different interpretations of what it is and what it might be.

(Massey 1994: 138; italics in original)

Not only is Massey dispelling the idea of stasis in terms of place here, she is also challenging understandings of place developed by previous regional and humanistic geographers that emphasise boundaries and conflate them with identities. Massey saw this previous local interpretation of place as one that enabled the construction of a singular narrative of place by a dominant majority that excluded those perceived not to belong. In the UK in particular, this type of narrative often relates to class, race, or both, and we can see Massey alluding to that in the different perceptions cited within the quote above. Place-attachment naturally alludes to ideas of belonging, which, in turn, inevitably links to notions of inside(r) and outside(r). In many of these dominant narratives visions of place are constructed specifically through a history of those who are seen as insiders. For example, May's (1996) research in Stoke Newington in the London Borough of Hackney was developed from ethnographic research with local residents. Stoke Newington is an area in East London where working class white residents live alongside large numbers of different ethnicities as well as recently arrived, predominantly white middle and upper class residents, many working within the media and cultural institutions, who have been drawn in by the process of gentrification. In interviews with different types of participants, different narratives were revealed. For example, one respondent reminisced about the times when one could leave one's door open without fear of intruders or burglary (May 1996: 200). Her view of Stoke Newington is that rather than thriving it is declining and this, in her view, is largely due to immigration. This perspective, and therefore her dominant, selective narrative, is formed by a sense of belonging and position as an insider, someone who has lived in Stoke Newington prior to gentrification and prior to the increased ethnic diversity of the population.

Conversely, Massey's progressive sense of place framed places as constituted as much by external forces as internal. This coming together of the global and local makes it difficult to establish what exists inside and outside of places, and therefore calls into question our understanding of boundaries. If a place is constructed through its global connections and flows, this suggests local boundaries are at least permeable, if not non-existent. However, returning to May's (1996) work, it is clear that not all views of place that refer to both global and local forces could be deemed progressive versions of place. For example, one respondent cited the diversity of Stoke Newington as one of its attractions (May 1996: 206). Whilst this initially seems to reflect the oft-repeated narrative associated with Hackney – that it is an area where multi-culturalism thrives and a diverse population lives together in relative harmony – there is a sense of an 'aestheticised difference' that is both enjoyed and viewed from a distance (Cresswell 2015: 112).

Alongside rethinking place, Massey's (2005) work has also reframed ideas of space, positioning it as active, rather than inert. To this extent, relational geographies have privileged space over place, perhaps not in the same way that previous quantitative, rational geographic approaches have, but space has been positioned as the active participant in the space-place relationship. This is evident when thinking about the global and the local as relational geographies propose that it is the effect of the global flows and networks that result in the particularity of place, therefore place is a product of these spatial flows and essentially passive in itself (Price 2013: 120). To this extent, interest in and theorising of place and place-making has diminished in geography, yet the richness of place and the vivid reality of everyday life that the humanistic geographers perceived is still evident in Massey's descriptions of place. What is also evident in both Tuan and Massey's writing about place is that the experience of one's everyday life in place is key to the construction of place and the meanings we assign to it. Place is continually being produced and reproduced by the repetition of one's everyday practices, even the most mundane or seemingly unimportant daily actions and interactions are an integral part of the process of place.

Everyday life

The concept of everyday life makes its first appearance in social thought in the 1920s, but it wasn't until after the Second World War that everyday life emerged as a recognised area of intellectual and artistic inquiry (Bennet & Watson 2002). The term everyday life generally refers to the lives of 'ordinary people', not the lives of the rich and powerful in the élite social classes, but those of the working and middle classes. The emergence of it as an area of study also stemmed from the democratisation of the term public, and the gradual expansion of what was thought to be worthy of public representation. Organisations such as Mass-Observation in the UK, writers such as Michel de Certeau (1984, 1998), Georges Perec (1997), and Henri Lefebvre (1991), and groups such as the Situationist International, have ensured that everyday life has since continued to be a productive site of inquiry throughout art and the social sciences. The concept of everyday life or the 'everyday' is resurgent in geography and is 'one of the key modes of thinking about contemporary urban geographies' (Yi'En 2014: 212). Similarly, like place, everyday life is not a static phenomenon, but is a 'dynamic process which is continually unfolding and emergent' (Eyles 1989: 102). The context in which everyday life is framed within this book is that of a 'taken for granted reality', a 'social construction which becomes a "structure" in itself': 'Everyday life is, therefore, the plausible social context and believable personal world within which we reside' (Eyles 1989: 103). Whilst this definition offers some certainty, inherent in it is a level of uncertainty. As with place, it is our own subjective experience that contributes to our understanding of everyday life. So, as with place, everyday life is 'characterised by ambiguities, instabilities and equivocation' (Highmore 2002: 17).

In a similar way to a relational geographic view of place as unbounded and open, a coming together or constellation of both global and local effects, the anthropologist Tim Ingold views place as a 'zone of entanglement' (Ingold 2008a,

2008b). For Ingold, this entanglement is essentially a 'meshwork', with specific sites within places, such as one's house or workplace, connected through a series of threads and knots (Ingold 2008a, 2008b). This conception of place evokes writer Italo Calvino's (1974) story of Ersilia, a place where residents connect their houses with strings that signify a particular relationship, for example, family, trade or authority. As place continues to evolve and relationships grow and extend, the strings become more numerous and the residents are unable to move through them. They therefore dismantle their houses and rebuild their city elsewhere. However, Calvino notes that the remaining meshwork of strings is still the city of Ersilia, even without the residents, who Calvino describes as 'nothing' (Calvino 1974: 76). For both Calvino and Ingold, place is constituted by these connections. For Pink, this conceptualisation of place 'offers a way of understanding practices as part of place' (Pink 2012: 25), that practices don't just 'happen *in* places', rather they are 'constituents of places'. Thus, the practice of everyday life is 'situated within a constantly changing constellation' of 'agencies, discourses, representations, materialities, persons, sensory and affective qualities, memories, texts and more' (Pink 2012: 28). Therefore, place and the practice of everyday life are inextricably interconnected, both contingent and mutually interdependent (Pink 2012: 29).

Concepts of practice have become central to contemporary discussions of everyday life, and like the spatial turn referred to previously, a 'practice turn' and a focus on analysing practices from a sociological perspective has emerged within a variety of disciplines (Pink 2012: 15–16). Many of these practices engage tacit knowledge or are undertaken as part of a regularly performed embodied routine within place, and this makes much of the everyday difficult to attend to as we may barely register many of its occurrences and therefore may see it as unimportant. To this end, many approaches to the everyday attempt to reverse this over-familiarity and find the extraordinary within the everyday. A further similarity in many projects that theorise or chart the everyday is the positioning of it as a site of resistance. This tactic of making the familiar strange is one that has its roots in the work of the Surrealists. Surrealism was an avant-garde movement established in the 1920s in Paris, drawing together a range of artists and writers. In contemporary art and design curricula, Surrealism is often positioned as an art movement and one that, for students and practitioners, supplies a convenient set of 'innovative techniques' (Highmore 2002: 46) for developing tangential ideas, freeing up imagination and the creative process, for example, in the *Exquisite Corpse* game in which a group of people each draw an element of a picture without being aware of what the previous people have contributed, resulting in a collaboratively constructed image. This type of tactic is rooted in the Surrealists' interest in the unconscious mind and the work of the psychoanalyst Sigmund Freud. However, Highmore prefers to position Surrealism as a 'continuation of avant-gardism in general' and suggests that it can be read as a 'form of social research into everyday life'. The products of Surrealist practices are therefore not art forms but 'documents of this social research'. Adopting this view, enables techniques such as collage to be used as 'methodologies for attending to the social' (Highmore 2002: 46). Collage and montage are practices which the Surrealists used regularly in order to juxtapose images and texts that were unexpected which therefore created a kind of 'charge' for the viewer and consequently resulted in the viewing of the collage as a new single image. For the Surrealists this was not simply a technique to artificially induce a sense of strangeness to the ordinary; their perception of the everyday was that it was 'collage-like' and strange already (Highmore 2002: 46). Much like Massey's description of Kilburn, and the conception of relational place generally, the Surrealists recognised 'the everyday as a montage of elements' (Highmore 2002: 47). The Surrealists' approach to the everyday influenced a range of artists, researchers and avant-garde movements whose work also centred on everyday life and place. These included Walter Benjamin, The Situationist International and Michel de Certeau. In turn, the writings and methods of these three proponents continue to influence contemporary approaches to everyday life and place. In particular, their engagement with walking, which has its roots in the practice of the flâneur, will be discussed in depth in Chapter 4.

Summary

Place, practice and the everyday have become key sites of study within geography and the social sciences more broadly. These interlinked, co-constitutive modes of thinking through and about urban geographies have resulted in something of a return to many of humanistic geography's traditions and have re-emphasised place as a site of study and an interest in methodological approaches that are 'inspired by resuscitated forms of phenomenology' (Yi'En 2014: 212). This type of approach, focusing on an embodied, experiential understanding of place is seen in both ethnographic approaches to place, which will be discussed in Chapter 4, and also in ideas of the non-representational which will be discussed in the following chapter. However, to conceive of both place and everyday life as continually emergent and always in process presents a methodological problem for those attempting to understand and represent examples of these two inextricable concepts. As a researcher of place and the everyday we are necessarily stepping into the flow of events, yet any attempt to arrest this flow in order to scrutinise it will be problematic (Highmore 2002: 21). However, the raison d'être of the scholar interrogating both place and everyday life is first to understand and then second to represent this understanding, and thus disseminate it to others. It is perhaps through the process of representation that place and everyday life is in danger of becoming a fixed, static entity, one that perhaps tells a singular narrative of a particular moment, frozen in time – entirely what place and the everyday aren't. However, as we will see in the following chapter, stasis might be an issue with a traditional form of geographic representation such as the map, but the theory and practice of representation has been challenged and developed in line with contemporary understandings of place and everyday life.

Bibliography

Agnew, J. (2005) 'Space : Place', in Cloke, P. & Johnston, R. (eds) *Spaces of Geographical Thought: Deconstructing Human Geography's Binaries*. London: Sage, pp. 81–96.

Anderson, J. (2010) *Understanding Cultural Geography: Places and Traces*. Abingdon: Routledge.

Augé, M. (1995) *Non-Place: Introduction to an Anthropology of Supermodernity*. London: Verso.

Bachelard, G. (1994) *The Poetics of Space*. Boston: Beacon Press.

Bennet, T. & Watson, D. (2002) *Understanding Everyday Life*. Oxford: Blackwell.

Calvino, I. (1974) *Invisible Cities*. San Diego: Harcourt Inc.

Castree, N. (2003) 'Place connections and boundaries in an interdependent world' in Holloway, S. L., Rice, S. P. & Valentine, G. (eds) *Key Concepts in Geography*. London: Sage, pp. 252–282.

Crang, M. (1998) *Cultural Geography*. London: Routledge.

Crang, P. (1999) 'Local-global' in Cloke, P., Crang, P. & Goodwin, M. (eds) *Introducing Human Geographies*. London: Arnold, pp 24–34.

Cresswell, T. (2015) *Place: An Introduction*. 2nd edition. Chichester: John Wiley & Sons.

Cresswell, T. (2013) *Geographic Thought: A Critical Introduction*. Chichester: John Wiley & Sons.

De Certeau, M. (1984) *The Practice of Everyday Life*. Berkeley: University of California Press.

De Certeau, M., Giard, L. & Mayal, P. (1998) *The Practice of Everyday Life Volume 2: Living and Cooking*. Minnesota: University of Minnesota Press.

Eyles, J. (1989) 'The geography of everyday life' in Gregory, D. & Walford, R. (eds) *Horizons in Human Geography*. London: Palgrave Macmillan, pp. 102–117.

Gieryn, T. F. (2000) 'A space for place in sociology', *Annual Review of Sociology*. 26, pp. 463–496.

Harvey, D. (1996) *Justice, Nature and the Geography of Difference*. Oxford: Blackwell.

Highmore, B. (2002) *Everyday Life and Cultural Theory: An Introduction*. London: Routledge.

Holloway L. and Hubbard, P. (2001) *People and Place: The Extraordinary Geographies of Everyday Life*. Harlow: Pearson Education Ltd.

Hubbard, P. & Kitchin, R. (2011) 'Introduction: Why key thinkers? in Hubbard, P. & Kitchin, R. (eds) *Key Thinkers on Space and Place*. 2nd edn. London: Sage, pp. 1–17.

Ingold, T. (2008a) 'Bindings against boundaries: entanglements of life in an open world', *Environment and Planning A*. 40(8), pp. 1796–1810.

Ingold, T. (2008b) 'Against Space: Place, Movement and Knowledge', in Winn Kirby, P. (ed.) (2009) *Boundless Worlds: An Anthropological Approach to Movement*. Oxford: Berghahn Books, pp. 29–43.

Latham, A. & McCormack, D. P. (2007) 'Developing 'real-world' methods in urban geography fieldwork', *Planet*. 18(1), pp. 25–27.

Lefebvre, H. (1991) *The Production of Space*. Oxford: Blackwell.

Massey, D. (2005) *For Space*. London: Sage.

Massey, D. (1994) *Space, Place and Gender*. Minneapolis: University of Minnesota Press.

May, J. (1996) 'Globalization and the politics of place: Place and identity in an inner London neighbourhood', *Transactions of the Institute of British Geographers*. 21(1), pp. 194–215.

Perec, G. (1997) *Species of Spaces*. London: Penguin.

Pink, S. (2012) *Situating Everyday Life*. London: Sage.

Price, P. L. (2013) 'Place' in Johnson, N. C., Schein, R. H. & Winders, J. (eds) *The Wiley-Blackwell Companion to Cultural Geography*. Chichester: John Wiley & Sons, pp 118–129.

Relph, E. (1976) *Place and Placelessness*. London: Pion.

Rose, G. (1993) *Feminism and Geography: The Limits of Geographical Knowledge*. Cambridge: Polity.

Tuan, Y. F. (2007) *Space and Place: The Perspective of Experience*, 25th anniversary edn, London: Edward Arnold.

Yi'En, C. (2014) 'Telling Stories of the City: Walking Ethnography, Affective Materialities, and Mobile Encounters', *Space and Culture*. 17(3), pp. 211–22.

2 Representing everyday life and place

Introduction

Ideas of the visual in geography are well established in terms of formal carto-graphic practices, the analysis of cultural practices such as landscape painting, and, perhaps more recently, imagery that geographers construct themselves as a way of representing place. Possibly the most iconic visual imagery associ-ated with geography is that of the map, which in taking a scientific approach to the recording of space positioned maps and mapping as neutral and objective. However, in the late 1980s, the map was challenged and revealed as subjective, rhetorical and power laden. Whilst this realisation offered the potential for see-ing maps as a form of agency, it failed to challenge a map's ontological security. However, more recent conceptions of maps have positioned them as 'processually emergent' and thus ontogenetic in nature (Kitchin & Dodge 2007).

Wider critiques of all types of representations were developed across the social sciences in the late 1980s, largely through the development of postmodernism which challenged notions of an absolute truth. Such metanarratives were there-fore revealed as reflecting a particular perspective – often that of the white, male academic – whilst silencing those from different standpoints. Poststructuralist work also contributed to these shifts, questioning and challenging the binaries that much of our understanding is built upon. Language was defined as inherently unstable, with meaning in a constant state of deferral. Therefore the metaphor of a text was often used to discuss place and landscape in relation to its un-fixed nature and its constant state of becoming. In the early 2000s, a non-representational approach to everyday life and place succeeded that of postmodern theory and built on previous poststructuralist approaches. Non-representational work endeavours to capture this continual state of emergence and does so through an embodied, experiential approach that draws from phenomenology. Therefore, while striv-ing to understand and represent a sense of everyday life and place, contemporary qualitative research does so in the knowledge that there is no single 'truth' waiting to be revealed. Yet, with regard to maps, this was not always the case.

Mapping place

When most people think of an image that relates to geography, it is quite likely they will think of a map. Whether it is a map that is centuries old which originally charted new lands far overseas, a large scale walking map complete with contours and marked trails, or a high school atlas, these varied visual forms are all synonymous with the idea of geography as a spatial science. That the geographic visualisation of space has been dominated by this scientific approach since the Enlightenment, has resulted in the map developing an air of neutral authority and unquestionable truth: 'No other image generated by human effort is granted such exemption from the personal, the subjective, the assumption of interestedness with which we automatically invest paintings and drawings (even photographs), essays and history (even eye-witness accounts)' (Wood 1992: 66). The 'foundational ontology' of traditional cartography is that objectivity and truth about the world can be achieved through the use of scientific techniques in the recording and communication of spatial information (Kitchin, Perkins & Dodge 2009: 12). The traditional map translates three-dimensional space into an unpeopled, two dimensional surface. It positions the user above and outside the territory, playing the 'god trick' (Haraway 1988), allowing one to take an omnipotent position, able to survey everything at once, with no hindrance of darkness, regardless of time of day or time zone (Pickles 2004: 80). The map gives no sense of the vertical, the emphasis is on the ground – suggesting 'the city is a place to be traversed . . . not an area to be lived in and through' (Black 1997: 13). As de Certeau (1984: 97) states, the 'geographical system' that transforms 'action into legibility . . . causes a way of being in the world to be forgotten'. This way of being is developed through our interaction with places and our movements. It is our own point of view and is therefore 'indexical'. The map, however, is non-indexical, offering the viewer a single generic view of place (Ingold 2000: 223). However, contemporary geographical thinkers have moved beyond this idea of a map as an impersonal 'truth'.

Deconstructing the map

Harley's (1989) seminal paper *Deconstructing the Map* sought to reposition understandings of the map away from the neutrality of science and towards a rhetorical device and representation of power. Harley suggested that those who ascribe to the scientific model of cartography believe that the route to cartographic objectivity and truth can be found through the continued refinement of systematic observation and measurement (Harley 1989: 2), yet he saw cartography as a form of language. Influenced by poststructuralist theorists and writers of the time, in particular Foucault and Derrida, Harley viewed maps as texts that could be read and consequently deconstructed 'to discover the silences and contradictions that challenge the apparent honesty of the image' (Black 1997: 18). This process of deconstruction ultimately severed the unquestioned link between reality and representation (Harley 1989: 3). Other scholars developed Harley's perspective, with the map being repositioned as an inscription (Pickles 2004) or a system of propositions (Wood & Fels 2008), rather than a representation. In this sense, the map '*creates* ideology, transforming the world *into* ideology, and by printing the world on paper *constructs the ideological*' (Wood & Fels 2008: 7; italics in original).

Although Harley's paper, and the many that followed it, led to the development of a critical cartographic movement that turned away from a positivist, scientific epistemology, his reading of Derrida and others generated some criticism. For example, Belyea (1992) accuses Harley of a superficial reading of the theorists behind his ideas. Derrida (1976: 158) states that 'there is nothing outside the text', which suggests that Harley had not really understood that this means the form of the map itself is devoid of meaning and has no fixed or objective relation to any kind of reality. In other words, Harley, whilst repositioning maps as social constructions, did not go as far as to challenge the map itself and question its onto-logical security – his position on maps suggests that 'a representational truth' is capable of being read if the ideological position of the map maker can be 'exposed and accounted for through deconstruction' (Kitchin, Gleeson & Dodge 2012: 2). So, the problem was not inherent in the map itself, but it was 'the bad things people *did* with maps' (Wood 1993: 50; italics in original). As we will discuss in the next section, following such poststructuralist representational arguments to a logical conclusion ends with the inability to produce or discuss representations at all. So, perhaps Harley was endeavouring to temper the abstract theoretical propositions in order to provide a more stable platform for the introduction of his socio-political agenda, and to persuade the discipline to engage with a 'postmod-ern climate of thought' (Harley 1989: 1). Wood describes this as letting fresh air into the overheated study (2002: 150) and legitimating a 'new discourse about maps' (2002: 156).

During the late twentieth century, and prior to Harley's deconstruction of the map, cartographers within the US sought to create a 'science of cartography' focusing on 'map effectiveness'. In doing so, their focus was on reducing error and developing graphic techniques that enabled the successful transmission and reception of information. Drawing on the work of Shannon & Weaver (1949), the aim of the cartographer was simply to 'reduce signal distortion in the communica-tion of data to users' (Kitchin, Perkins & Dodge 2009: 5). Today the discipline of cartography continues to develop in relation to how best to display spatial infor-mation and focuses on issues such as accuracy and readability (Kitchin, Gleeson & Dodge 2012: 1–2). It remains mostly untouched by contemporary theoretical debates, and similar divides persist between cultural geographers and cartogra-phers; science and art; and image and text. As a cartographic historian, Harley was concerned with developing a theoretical, socio-political framework for the reading of maps, and offers few strategies to change the practice of mapping. It is easy to see how the divide between theorists and practitioners developed, with cartographers accused of losing interest in 'the meaning of what they represent (Harley 1990: 7), and of being mere 'manipulators and generalisers of other peo-ple's data' (Harley 1990: 10). These criticisms are partly due to cartographers seeking validation of their practice under the auspices of science – they gain aca-demic status, but the subject is reduced to the development of technology and aesthetics with no involvement in content or meaning (Harley 1989: 10).

More often than not, binary opposites such as art and science are used to criti-cise and demean each other, or are used unproductively, to define the discipline in a negative way – 'cartography is an art because it's not a science' or 'cartography is a science because it is not an art' (Krygier 1995: 4). Science's domination has been established through two particular arguments. First, that the graphics on a map are simply the outcome of a scientific process and their value is not linked to art. This relies on the definition of science as objective and neutral and art as subjective and indulgent. Second, that scientific investigation is purely about progress, with new knowledge replacing old; unlike art, which sits old masters alongside contemporary artists in a museum (Krygier 1995: 6). Blakemore & Harley (1980: 13) suggest that there is a need to move towards an articulation of 'a fundamental principle, deeper than form or content'.

The agency of mapping and maps as narratives

A fusion of both form and content can be seen in work that endeavours to use the potential of mapping for liberating or productive ends and to tell particular stories. Much of mapping's roots is in authoritarian exercises undertaken by those colonising, exploring or conquering new territories. For example, the British Ordnance Survey maps – probably the type of map most used in the UK by those who are interested in leisure pursuits such as walking, cycling or running in the countryside – have their history in military strategy and the mapping of the Scottish Highlands after the rebellion of 1745. Therefore, as we have seen above, there is no doubt that maps are socially constructed and can be used to coerce or control. For example, the choice of what is centred or what is at the margins of the frame can indicate how the creator wishes to position what is of importance, or what might be seen as 'self' and 'other' in a way to define a collective identity. This is not a new phenomenon, ancient maps often represented unknown people living in strange lands as monsters so as to define a boundary between the known and unknown.

However, it is possible to reposition the idea of mapping as a 'collective enabling enterprise', a process that has agency and is able to reveal 'hidden potential'. In this context, rather than project a dominant narrative of the powerful few, mapping can be used as an emancipatory force, enabling a diversifying and enriching of our understanding of place (Corner 1999: 89). So rather than see accuracy or imposition as mapping's most productive effects, Corner suggests that if mapping is thought of and used as a creative practice it can uncover previously unseen or unimagined realities. Whilst Corner's focus in this particular article is the potential of re-envisioning mapping in the context of design and planning, it is clear that mapping, particularly at a time where there is a rise in interest in data visualisation generally, can be used to tell stories from a minority perspective and to shift the perceived edges of the map to the centre.

In order to develop a map that possesses this kind of agency, the creator needs to go beyond simply reproducing what is already known about the territory, to focus on elements beyond the terrain or topography. Deleuze and Guattari (1987: 12) define maps that simply attempt to mimetically reproduce space as 'tracings', suggesting that this is about 'alleged competence'. Alternatively, they define a map as being centred on the idea of 'performance'. It is this performance that enables the map to repeatedly re-make the territory and enable 'new eidetic and physical worlds to emerge' (Deleuze and Guattari 1987: 12). By including a diverse range of other elements such as local narratives, historical events, environmental factors or political interests, the map shows that the physical terrain is the 'surface expression of a complex and dynamic imbroglio of social and natural processes' (Corner 1999: 90). In highlighting this interrelationship, the map has the potential to become 'an active agent of cultural intervention' that visualises the world in new ways that may result in change (Corner 1999: 91). This approach to mapping has its roots in Harley's work and that of other critical cartographers. Critical cartography, focuses on the political analysis of maps and mapping practices but is not a movement against maps. Rather, critical cartographers 'appreciate the diverse ways in which maps are produced and used' and therefore, do not subscribe to a 'right way' of producing maps. What they do require of map makers is a sensitivity to the politics and context of any map's making and use (Kitchin, Perkins & Dodge 2009: 10).

The work of Harley and others in exposing and deconstructing the metanarratives inherent in cartographic practices has enabled maps to be proactively envisioned as a vehicle for storytelling (Caquard 2011: 136). This has coincided with a rise in interest in spatiality within art and design (see for example, Harmon 2004; Abrams & Hall 2006) and has led to the reinvigoration and reframing of the traditional practice of mapping. DeLyser & Sui (2014: 296) note that by the end of the 1990s, the use of maps and mapping had declined within human geography – perhaps not surprising considering the critiques of mapping in the late 1980s discussed above, and the 'crisis of representation' which will be discussed in the following section. Yet, in recent years, the map has risen in popularity and has been 'reinvigorated' as a research method (DeLyser & Sui 2014: 296). Whilst this is partly due to new technological opportunities such as geographic information systems (GIS), the Global Positioning System (GPS) and open source mapping software, it is not solely down to this. Since the critical turn in cartography, interest in the potential of exploring narratives via the practice of mapping has grown. This interest in narrative or storytelling as a method is increasingly popular across the discipline as a whole and aligns with 'the relational and material turn in geography' (Cameron 2012). In the context of mapping specifically, terms such as 'story map', 'fictional cartography', and 'geospatial storytelling' have emerged and the interest in mapping as a process or tool used in the development and communication of a narrative is also evident within the social sciences and the arts – as is an emerging interest in spatiality (Caquard 2011: 135). A story map is a form of spatial expression that embodies personal experience of a particular environment which thus contributes to a deeper understanding of place (Caquard 2011: 136).

The potential of maps to move beyond the flatlands of the traditional gridded map we find in the pages of a road atlas has been explored by many artists and authors in recent years. For example, artist Christian Nold creates maps that often focus on one's emotional response to the urban landscape. He looks to populate his work, as people are central to cities, yet rarely represented on traditional maps. Nold's *Bio Mapping* (2004) is a participatory methodology that uses technology to record an emotional response to place. Participants wear a galvanic skin response sensor that measures sweat levels, which is linked to a GPS unit, enabling Nold to pinpoint changes in the body's skin conductivity in specific geographical locations. On returning from their walk, participants are asked to reflect on these differing areas of excitement or stress and try to remember what triggered this response. This text is then included within the graphical representations. The representations themselves also vary, with some, *The Stockport Emotion Map* (Nold 2007), for example, also including participants' drawings in response to questions in relation to their feelings about aspects of the town. These maps generate a montage of feelings, experiences, sights and sounds that require the viewer to piece together a series of narratives about place.

Whilst Nold does use technology as part of his mapping practice, several writers caution against methodological novelty which may obscure or distract from engaging with the actual issues of the research (see for example DeLysier & Sui 2014: 297; Travers 2009). Traditional forms of mapping undertaken both as data gathering or disseminating therefore endure, and rather than see this as a return to methods that were previously thought to be deceased or on the wane, DeLyser & Sui (2014: 297) suggest it is 'a recognition of the continued relevance' of mapping 'to the critical questions of our changing world'.

Processually emergent mappings

Recent work on cartography has reframed it as a processual rather than representational science and suggested that 'maps emerge through practices' (Kitchin & Dodge 2007: 331). Kitchin & Dodge's (2007) preferred term is 'mappings' and this 'processually emergent' understanding (Latham & McCormack 2004) questions the taken for granted ontological position of cartography as objective and truthful. It positions maps as 'ontogenetic in nature . . . always remade every time they are engaged with' (Kitchin & Dodge 2007: 335). Without these practices a map is simply a collection of lines, points and ink shaded areas – the map reader has to bring it to life. This constant usage contributes to the belief of 'ontological security' as the user's knowledge of map use develops and is reaffirmed each time (Kitchin & Dodge 2007: 335). This sense that maps are remade each time they are engaged with suggests that they are not simply fixed representations of place, but can be used and interpreted in multiple ways.

> [M]aps stretch beyond their physical boundaries; they are not limited by the paper on which they are printed or the wall upon which they might be scrawled. Each crease, fold, and tear produces a new rendering, a new possibility, a new (re)presentation, a new moment of production and consumption, authoring and reading, objectification and subjectification, representation and practice.
>
> (Del Casino & Hanna 2006: 36)

There is a resonance here with Barthes' (1977) notion of the 'death of the author', with the map being produced and reproduced by the reader. Del Casino & Hanna's move is towards 'map spaces' that are simultaneously both representations and practices, which make the non-productive binary oppositions of text and space, experience and representation redundant (Del Casino & Hanna 2006).

Maps as perceived by Del Casino & Hanna (2006: 37) are 'tactile, olfactory, sensed objects/subjects mediated by the multiplicity of knowledges we bring to and take from them'. As such, they have the potential to become, and to hold, biographical information. In Harley's (1987) discussion about a map of Newton Abbot, a town in Devon in the UK, which he has taken with him when emigrating to the US, he suggests it 'restores time to memory and it recreates for the inner eye the fabric and seasons of a former life' (Harley 1987: 330). When Harley looks at his map on the wall in his new home he remembers his English roots, the schools his children attended, the church his daughter was married in, and the churchyard his wife and son are buried in. That all this can be conjured from a series of lines, points, shadings and colours, confirms that the map has the potential to reverberate beyond a replication of the terrain, and has the potential to act 'not so much as a topography as an autobiography' (Harley 1987: 330). This way of thinking, and that of both Kitchin & Dodge and Del Casino & Hanna inextricably link the map with place and the experience of that place, and enable the possibility of grounding a representation of place in print without 'fixing' the meaning. Maps do not re-present or make the world, rather they are 'a co-constitutive production between inscription, individual and world' (Kitchin, Perkins & Dodge 2009: 21). This offers us an opportunity to rethink maps and other visual representations as less about 'truth' and more about experience, thus avoiding many of the challenges brought about by the 'crisis of representation'.

Visual representation and reality

In the late 1980s the social sciences, and by extension cultural geography, were shaken by the 'crisis of representation'. This representational challenge was brought about by the advent of postmodernism which challenged the understanding of, and approaches to, representation. Postmodernist thought challenged the notions of an absolute 'truth' which had been a central tenet of Modernist thought, developed since the Enlightenment in the eighteenth century. The roots of the Enlightenment are found in the scientific revolution of the sixteenth and seventeenth centuries, which paved the way for a belief that it is possible to understand and explain the world and thus reveal its true order. In this context, and as the cartographers discussed above and the humanistic geographers discussed in the previous chapter believed, it is simply a question of progress in relation to science or philosophy that will inevitably result in 'better, more truthful understandings of people and place over time' (Holloway & Hubbard 2001: 283). Modernists of all guises therefore believed that rational thought and an objective, scientific approach could provide an incontrovertible answer to life's problems and questions – which essentially equates to a positivist approach. To develop what might be considered the absolute understanding of a particular place in these terms would involve ensuring spatial data is mapped as precisely as possible, numbers of residents are recorded accurately, and particular types of spaces are defined clearly, or that any description of events is as clear as possible. This in turn leads to the development of a single metanarrative – one overarching theory that can explain the complexities inherent in a particular place. However, this first assumes that there is a conception of place that is experienced by everyone. A postmodern philosophy eschews assumptions such as homogeneity and certainty, and is underpinned by ideas of heterogeneity, relativity and difference. Thus, a postmodern approach challenges the very possibility of 'truth' being established via 'accurate description, rational thought or representational accuracy' (Holloway & Hubbard 2001: 283).

Second, it assumes that language and other representational strategies have an inherent transparency, and relate directly to the thing that was being represented. Postmodernism challenged this, suggesting that the representations simply referred to further representations, or in semiotic terms, that signs refer to other signs not to any kind of external reality. This effectively severed the 'supposed one-to-one link between language and brute reality' (Barnes & Duncan 1992: 2). It also led postmodernists to the conclusion that representations produce reality and play a role in constructing place as we understand and experience it – that language and other representational strategies are therefore neither inactive nor neutral. Philosopher Jean Baudrillard (1983; 1994) takes this idea to the extreme in his writings on simulation and the hyper-real.

> Simulation is no longer that of a territory, a referential being or a substance. It is the generation by models of a real without origin or reality: a hyper-real. The territory no longer precedes the map, nor survives it. Henceforth, it is the map that precedes the territory. . .
>
> (Baudrillard 1983: 2)

According to Baudrillard, the signs we see and read in the media or popular culture no longer refer to any kind of reality. Therefore, in a postmodern context the simulacra replace the original.

Less extreme than Baudrillard, but echoing discussion in the previous chapter regarding a 'denial of difference', Shields' (1991) concept of 'place image' highlights the very real link between 'reality' and representation. Place-image is defined as the development of an oversimplified narrative that plays a part in constructing place by affecting one's decision as to whether to visit somewhere. Such place-images emerge in three ways; through 'over-simplification' – reducing the place to a singular trait; heightening the focus on one or more specific traits; and labelling – defining a place to be of a certain nature. Such traces are likely to remain in culturally produced artefacts even after the nature of the place has changed (Shields 1991: 47). That representations play a role in constructing place leads inevitably to the conclusion that different 'truths' are likely to be developed depending on the different standpoints of the authors. The choice of which story or stories one tells is therefore an inherently political and moral decision (Holloway & Hubbard 2001: 248). Whilst this applies to postmodernist thinking, feminist geographers had previously developed a similar proposition, revealing that knowledge is always situated, always reflecting particular 'perspectives, political biases, and cultural values' (Cresswell 2013: 158). This was particularly key in relation to critiquing the white, male perspective that had been the dominant voice within geography, as with many other academic disciplines.

Poststructuralist ideas have also had a profound effect on our understandings of place, reality and truth. Often used interchangeably with postmodernism, and with some evident similarities, poststructuralists believe that underlying generative structures, which allegedly offer clarity about how everyday life and place emerge, are impossible to identify (Cresswell 2013: 207). Perhaps inherent in humans is a desire to make sense of things, and we often ascribe things as consequences of the particular categories in the process of doing this. We might, therefore, describe something like the Blue Mountains National Park in New South Wales, Australia as nature, for example. This enables us to classify and categorise in a way that creates boundaries and an illusion of clarity. However, it doesn't 'do justice to the messiness of the world' (Cresswell 2013: 207). For example, digging a little deeper into the description of the Blue Mountains National Park as nature would lead us to questions of how 'managed' nature has to be to remain 'natural' in the context of a public park that has to manage a large flow of tourists and keep footpaths and boundaries maintained in the face of natural forces such as rain, wind and fire that may lead to erosion over time.

Poststructuralism does not suggest we ignore such categories, but asks us to question how such categories are constructed, who has the power to construct them and how they function in everyday life (Dixon & Jones 1998: 254). A poststructuralist approach starts from questioning the binary opposition of such terms as nature/culture, place/space and local/global as the terms are seen as relational – one does not exist or have meaning without the other. It assumes that reality is continually being produced and reproduced on the 'surface' of place by these sets of relations (Cresswell 2013: 208). Rather than clearly defined places, poststructuralism also 'tends to think in terms of flows, networks, and folds' (Cresswell 2013: 215) which brings us back to Massey's definition of place and its complex and contradictory nature that results in 'loose ends' (Massey 1997: 222) and a 'sieve order' (de Certeau 1984: 107). These types of descriptions, and the above discussions as to the issues with representation itself, suggest that to represent place is all but impossible. However, some geographers have used this postmodern critique of representation as a way to rethink how we might use the idea of the 'text' to analyse place.

Landscape as text

A postmodern understanding of a 'text' has been proposed by Barthes, amongst others. In these terms a text is not simply a piece of writing that resides on a printed page, it encompasses all culturally produced items, such as maps, paintings, buildings and landscapes. Since the 'crisis of representation', language is no longer seen to mirror reality, rather it reflects and refracts our own experience as writers, and meaning is produced via the process of intertextuality in which texts draw on other texts, which in turn draw on further texts, and so on and so forth. Thus texts are 'constitutive of reality' (Barnes & Duncan 1992: 5). Barnes and Duncan suggest that in analysing landscapes the metaphor of a text is therefore appropriate as 'it conveys the inherent instability of meaning, fragmentation or absence of integrity, lack of authorial control, polyvocality and irretrievable social contradictions' (Barnes & Duncan 1992: 5). Similarly, Barthes' (1971: 168) poststructuralist view of the city positioned it as a discourse − 'the city speaks to its inhabitants, we speak to our city . . . simply by living in it'. In earlier structuralist works Barthes sought to 'decipher society's signs and to reveal the complexity and instability' of everyday life and place (Duncan & Duncan 1992: 18). However, his later poststructuralist work argues that a definitive interpretation of place is impossible, because of the inherent instability of meaning and the endless chain of signifiers and signifieds (Duncan & Duncan 1992: 26). Barthes suggests developing a multiplicity of readings of the city, generated by a range of readers, 'from the native to the stranger' (Barthes 1971: 164). Although Barthes moved from a structuralist position to a poststructuralist, in some ways he didn't seem to entirely disown his earlier realist ontological position (Duncan & Duncan 1992: 20) and ends *Semiology and the Urban* by suggesting that 'many of us should try and decipher the city we are in, starting if necessary with a personal rapport' (Barthes 1971: 164). The word 'decipher' seems to point to some kind of code that is discoverable, and perhaps contradicts his poststructuralist position of meaning as 'never fixed, but always in a state of deferral' (Leach 2002: 4).

Given this continual deferral of meaning, if one were to take the notion of landscape as text literally, then attempting to represent place becomes paradoxical. Daniels & Cosgrove (1988: 8) take such a poststructuralist view to its ultimate conclusion by suggesting that:

> From such a postmodern perspective landscape seems less like a palimpsest whose 'real' or 'authentic' meanings can somehow be recovered with the correct technique, theories or ideologies, than a flickering text displayed on the word processor screen whose meaning can be created, extended, altered, elaborated and finally obliterated by the merest touch of a button.

Their analogy is clear, however, as in reality it is difficult to fully delete a digital file. Like the palimpsest, a ghostly imprint of the data remains on the hard drive. In terms of place, long disappeared buildings or stories of past events do remain, and are referred to both formally and informally. Blue plaques reveal famous occupants of houses, memorials (both formal and informal) remember those who have lost their lives in an event, and directions given by those who are long term residents of a place can often refer to landmarks that no longer exist − 'take a left where the Star Garage used to be'. Ephemeral narratives or myths are also constructed around other places such as 'unusual houses, cemeteries, and lonely bridges' (Bird 2002: 525). These local narratives define a community in a certain way, highlighting outsiders or perceived transgressions of a particular value system. They tell us less about formal history and more about a community's need to 'construct a sense of place and cultural identity' (Bird 2002: 526). Representation seems to be an innate instinct then, and as we can see from the narratives discussed above, we spend much of our time telling stories – for example, recounting stories of our day to friends, or sharing memories with family (Laurier & Philo 2006: 354–355). So within geography, or in life generally, we are unlikely to resist representation, regardless of any kind of 'crisis'.

Highmore suggests that there may be 'forms of representation that are *more* appropriate, *more* adequate for attending to the everyday' and that to assume it is impossible to represent everyday life and place is to 'condemn it to silence' (Highmore 2002: 21: italics in original). Yet he also suggests that because everyday life and place are continually in process, to attempt to represent it is in effect halting this process and creating an artificial stasis through the attention given to one aspect. Therefore, everyday life and place will always exceed any attempts 'to apprehend' them. However, instead of accepting defeat in our attempts to develop representational strategies that reflect everyday life and place, Highmore suggests we should just accept that different representational forms are going to produce different versions or readings of everyday life and place (Highmore 2002: 21). Similarly, Cosgrove & Domash (1993: 35–36) suggest that 'The problem of representation is only a 'crisis if we somehow think that we are conveying some independent truth about the world, that we are relaying an authentic representation'. As with Harley's critical cartography, this statement seems to ignore the fact that it is representation itself that is at stake here. Cosgrove & Domash avoid engaging with postmodern thinking around language and representation. However, since then, non-representational theory, which succeeds postmodern theory and is a logical extension of poststructuralist thought, has challenged the triumvirate of understanding, meaning and the use of language (Vannini 2015: 2).

Non-representational theory

Non-representational theory, or 'more than representational' theory (Lorimer 2005) as it is often referred to, has recently become one of the 'most influential theoretical perspectives within social and cultural theory' (Vannini 2015: 2). Non-representational theory is an 'umbrella term' for a range of diverse work that attempts to contend with, and convey, our 'more-than-human, more-than textual, multisensual worlds' (Lorimer 2005: 83). Non-representational theory draws from a diverse range of disciplines including performance studies, cultural studies, the sociology of the body and emotions, and the sociology and anthropology of the senses. It also brings together a range of theoretical positions such as actor-network theory, performance theory, poststructuralist feminism, critical theory and postphenomenology (Vannini 2015: 3). It is perhaps the driving force behind the upsurge in work noted by Hawkins that pursues a range of experimental methods which 'situate the sensing body front and centre' (2017: 65).

Nigel Thrift, the key proponent of non-representational theory in the context of cultural geography, positions it as a 'series of procedures and techniques of expression' that result in 'an outline of the art of producing a permanent supplement to the ordinary, a sacrament for the everyday, a hymn to the superfluous' (Thrift 2008: 2). The raison d'être of non-representational theory is to attempt to 'capture the "onflow" . . . of everyday life' (Thrift 2008: 5), it is a 'geography of what happens' (Thrift 2008: 2) and sees the world as always in a 'state of becoming' (Creswell 2013: 227). In capturing this becoming, non-representational theory focuses on practices; 'material bodies of work or styles that have gained enough stability over time . . . to reproduce themselves' (Thrift 2008: 8). This prioritises the embodied and the performative, rather than representation and meaning (Thrift, 1997: 126–127). Lorimer (2005: 83) describes non-representational theory as centring on 'everyday routines, fleeting encounters, embodied movements, precognitive triggers, practical skills, affective intensities, enduring urges, unexceptional interactions and sensuous dispositions', all of which make 'a critical difference to our experience of . . . place'. These are the kind of experiences that happen without us sometimes even being aware of them, things we may experience or do without really knowing why, and things that we might view as unimportant or not worth reflecting on.

The vast majority of academic research output about everyday life and place is constructed representationally via textual means and, for non-representational theorists, these types of representation privilege the text rather than the experience (Nash 2000). In an attempt to move away from textuality, non-representational theory is also deliberately experimental, and there is no defined methodology nor any prescribed methods. As Vannini (2015: 5) puts it, 'non-representational work aims to rupture, unsettle, animate, and reverberate rather than report and represent'. Cultural geographers have begun to explore methods that utilise media such as film (see Lorimer 2010; Garrett 2011) and sound (see Gallagher & Prior 2014), as a way of engaging 'with the experience of the world in their moments of creation' (Anderson 2010: 32). As Marks has stated in relation to cinematic images, 'memory functions multisensorially and activates a memory that necessarily involves all the senses' (Marks 2000: 22).

Multi-sensory experience and sensory memories

The work of Merleau-Ponty is key to the development of a multi-sensory approach as he 'placed sensation at the centre of human perception' (Pink 2015: 29). One of Merleau-Ponty's propositions that has been developed by a range of writers, including Ingold (2000) is that our senses are interconnected; that we don't see or hear in isolation, but that our body is a 'synergic system all of the functions of which are exercised and linked together in the general action of being in the world' (Merleau-Ponty 2002: 272). Similarly, Ingold draws on the work of ecological psychologist James Gibson and the different modalities of sensory experience in suggesting that 'Looking, listening and touching, therefore, are not separate activities, they are just different facets of the same activity: that of the whole organism in its environment' (Ingold 2000: 261).

Not only are the senses interrelated and cross-modally active, they are also inextricably linked with memory. Proust (2013: 53), for example, famously writes of his involuntary memory triggered by the taste of a madeleine dipped in tea, which takes him back to childhood moments with his aunt. Anthropologist David Sutton (2001: 12) states that 'these types of memories can be found sedimented in the body' and that the sensuality of food in particular means it is a compelling channel for memories – 'the experience of food evokes recollection, which is not simply cognitive but also emotional and physical' (Holtzman 2006: 365). The idea of sedimentation suggests that these embodied memories build up through repeated practices, much like rock strata that have been deposited over time, reflecting the conditions that each layer was deposited and formed under. Thus it isn't solely our sensory environment that plays a part in this, but also our embodied practices – that, as Bachelard has also suggested, our memories are emplaced (Bachelard 1994: xxxii). Seremetakis (1994) similarly posits that meaningful objects carry 'emotional and historical sedimentation that can provoke' us into various types of action. However, rather than see these sensory memories and associated meanings as fixed' they can be viewed as 'continually reconstituted through practice' (Pink 2015: 44). So, rather than the memory repeating itself, it is made anew each time in the present (Seremetakis 1994: 7). Thus, in this context, one could perhaps view the rock strata as a single memory, being continually re-deposited over time under slightly different conditions. Readers will therefore undoubtedly bring to bear their own 'sensory memories' (Seremetakis 1994: 6–7) on a representation that has some form of resonance with them. Yet, although these memories can be likened to language, they cannot be reduced to language (Seremetakis 1994: 6), so in this context, it would seem that non-representational work might be able to sidestep verbal or written language.

Similarly, Crouch (2010) discusses the performative nature of two dimensional artworks through the viewer's engagement.

> The performative 'life' or vitality of the artwork—even two dimensional work—is performed too by the individual in his and her encounter with it. Two dimensional pictures may not be experienced only through the gaze, but with diverse dispositions of the body, memory, recall, intersubjectivity, emotion, fear and anxiety. . .
>
> (Crouch 2010: 8)

These 'sensory memories' and 'dispositions of the body' could be defined as affective responses, and affect is a key concept in non-representational theory. It is a difficult term to define, and it is often imprecisely articulated in terms of feelings or emotions. Vannini (2015: 8–9) defines affect variously as 'a pull and a push, an intensity of feeling, a sensation, a passion, an atmosphere, an urge, a mood, a drive—all of the above and none of the above in particular'. However, while there may be a relation between affect and emotion, they are not one and the same – the key difference is that affect comes before emotion. Emotions are essentially the way our bodies make sense of affect and give it meaning, and this meaning is inevitably determined both socially and culturally (Cresswell 2013: 230). If we are watching a horror movie alone in the dark and the hairs stand up on the back of our neck, we articulate this experience as fear. As soon as we verbalise the emotion we enter representational territory – affect is therefore seen to be pre-representational (Cresswell 2013: 230).

Non-representational methods

Although geographers have experimented with methods such as film and sound in order engage with the embodied, affective experience of the world, according to Vannini there is no single prescribed method for non-representational research. Rather, non-representational researchers use a range of methods – many of which are used by researchers undertaking research within other theoretical traditions – for example, interviews, participant observation, archival research and artistic interventions (Vannini 2015: 11). Similarly, there is also no 'unique non-representational mode or medium of communication', and non-representational research unfolds through 'writing, through photography, through dance, or through poetry, video, sound, art installations or any of the other research communication modes and media available in the twenty-first century' (Vannini 2015: 11). Whilst this may be surprising to some as non-representational theory is often discussed in the context of performance, Vannini suggests that performance is not a magical strategy, but has its own pros and cons. Yes, it is embodied, lively and relational, but it isn't always appropriate and is limited in its analytical depth (Vannini 2015: 11). What Vannini suggests is the key difference in non-representational solutions to the crisis of representation is a concern with 'issuing forth novel reverberations' rather than 'representing lifeworlds'. In effect, there is less of an interest in extracting or representing some kind of empirical reality from data, and more of an interest in 'enacting multiple and diverse potentials of what knowledge can become afterwards' (Vannini 2015: 12). In attempting to capture a sense of fleeting, embodied moments in the world, we lose 'immediacy', as 'physical presence and immediacy cannot be stored (in time) directly as representations' (Dirksmeier & Helbrecht 2008: 7). The problem with text-based versions of events is that 'the words try and catch up with the event . . . but . . . they always remain "afterwords"' (Dirksmeier & Helbrecht 2008: 10). So, rather than attempt to mimetically reproduce events of the past, non-representational work is more concerned with 'evoking, in the present moment, a future impression' on a reader or viewer or listener (Vannini 2015: 12). This positions non-representational work in a proactive space where the focus is on creating some kind of change or effect. These types of changes or effects can be wide ranging and might include some kind of social change, triggering an intellectual interest, challenging expectations, or generating new stories (Vannini 2015: 12).

Non-representational theory is not without its critics, and the turn from visual and literary texts to a focus on expressive, body-practices simply creates a new version of an old divide – theory and practice (Nash 2000: 657). Paradoxically, it is also impossible to engage with non-representational theory without engaging in representation. A film of a dance performance or a sound recording of a political rally is always once removed from the experience itself – it is a re-presentation just as it would be if it were textual. When these recordings are shown in exhibitions or at conferences, the audience also becomes an active participant in the construction of the re-presentation. We inevitably look for meaning in such contexts and, in doing so, we articulate our understanding of this meaning in thoughts and language. Sometimes these thoughts are triggered by a 'sensory memory' that is an embodied, affective response, but it is unlikely that these sensory memories are experienced in isolation. They are inevitably followed by thoughts of prior experiences and memories of certain people or places.

However, what non-representational theory does offer is a challenge to the 'methodological conservatism' (Latham 2003: 72) and 'timidity' (Thrift 2000: 3) of traditionally written up ethnographic research. We perhaps return to issues relating to traditional academic publishing at this point, with formulaic structures and formats impinging on these experimental possibilities and the 'hegemony of timidity' asserting 'its conservative power' (Vannini 2015: 13). Whilst for some this seems to have been taken as the need to reject traditional methods, as we have seen, non-representational work is produced in a variety of modes, writing included. So, rather than this being seen as a comment on the use of specific methods it should be taken in the context of reinterpreting and re-appropriating established methodologies. This approach, according to Latham, can be 'relatively modest' in terms of its experimentation, and benefits from 'the allowance of a certain amount of methodological naiveté'. It also requires a 'broadminded openness' in relation to methodologies and pluralism within the discipline of human geography (Latham 2003: 75). So,

> . . . rather than ditching the methodological skills that human geography has so painfully accumulated, we should work through how we can imbue traditional research methodologies with a sense of the creative, the practical, and being with practice-ness that Thrift is seeking. Pushed in the appropriate direction there is no reason why these methods cannot be made to dance a little.
>
> (Latham 2003: 72)

Whilst one can understand the general point Latham is making, phrases like 'pushed in the appropriate direction' or 'made to dance a little' provide little in the way of specificity. Yet, perhaps that is his point, rather than reinforce the mistaken idea that non-representational work must solely focus on the performative, he deliberately leaves the door open for a more inclusive methodological approach. However, the idea of methodological naiveté is also interesting in this context, and that of interdisciplinary work generally. If one is less familiar with the traditions of a discipline one could be said to be naïve by definition. In this sense, an interdisciplinary approach offers the opportunity to challenge assumptions or transgress perceived boundaries in a way that is difficult if one is steeped in a particular tradition. From my perspective, geo/graphic work offers an opportunity to do this. As someone who was not trained as a geographer or anthropologist, I can perhaps more easily appropriate theories, methodologies and methods in a way that makes sense within the context of geo/graphic research.

Summary

Representational strategies and critiques have been developed in such a way as to take account of contemporary definitions of place that position it as ongoing and relational and non-representational work contributes to the 'practice turn' identified within the previous chapter. Many of the approaches taken with regard to framing discussions of everyday life and place – and the representation of them – are interdisciplinary in nature, and as we will see in Chapter 4 this is also true of a geo/graphic approach. In terms of addressing accusations of methodological timidity, geographers have begun to engage with everyday life and place using methods that can be described as creative, more often than not drawing from artistic practices such as performance, poetry or other forms of creative writing. However, even though much of this work has been done in collaboration with artists and academics from different disciplines, to date there has been little engagement with the discipline of graphic design. Perhaps this is because the perception of design is that it is a service industry, existing primarily within the commercial context, incapable of offering the creative insights that art might. However, designers are used to bridging the divide between form and content and thus are well placed to contribute to, and collaborate on, such methodological developments. The following chapter discusses geographers' approaches to collaboration and creative methods to date, many of which have sought to eschew a traditional text-based approach and explore media that are perceived as less fixed in nature, such as film or sound. It is therefore followed by a discussion of graphic design that positions it as capable of creating text-based representations of place that, much like processually emergent conceptions of mapping, engage the reader in the re/construction of everyday life and place.

Bibliography

Abrams, J. & Hall, P. (eds) (2006) *Else/Where: Mapping New Cartographies of Networks and Territories*. Minneapolis: University of Minnesota Design Institute.

Anderson, J. (2010) *Understanding Cultural Geography: Places and Traces*. Abingdon: Routledge.

Bachelard, G. (1994) *The Poetics of Space*. Boston: Beacon Press.

Barnes, T. J. & Duncan, J. S. (1992) 'Introduction: Writing words' in Barnes, T. J. & Duncan, J. S. (eds) *Writing Worlds: Discourse, Texts, and Metaphors in the Representation of Landscape*. London: Routledge, pp. 1–17.

Barthes, R. (1977) *Image, Music, Text*. London: Fontana.

Barthes, R. (1971) 'Semiology and the urban' in Leach, N. (ed.) (1997) *Rethinking Architecture: A Reader in Cultural Theory*. Abingdon: Routledge, pp. 158–164.

Baudrillard, J. (1994) *Simulacra and Simulations*. Michigan: University of Michigan Press.

Baudrillard, J. (1983) *Simulations*. New York: Semiotext(e).

Belyea, B. (1992) 'Images of power: Derrida/Foucault/Harley', *Cartographica*. 29(2), pp 1–9.

Bird, S. E. (2002) 'It makes sense to us: Cultural identity in local legends of place', *Journal of Contemporary Ethnography*. 31(5), pp. 519–547.

Black, J. (1997) *Maps and Politics*. London: Reaktion Books.

Blakemore, M. & Harley, J. B. (1980) 'Definitions', *Cartographica*. 17(4), pp. 5–13.

Cameron, E. (2012) 'New geographies of story and storytelling', *Progress in Human Geography*. 36(5), pp. 573–592.

Caquard, S. (2011) 'Cartography I: Mapping narrative cartography', *Progress in Human Geography*. 37(1), pp. 135–144.

Corner, J. (1999) 'The agency of mapping: Speculation, critique and invention' in Dodge, M., Kitchen, R. & Perkins, C. (eds) (2011) *The Map Reader: Theories of Mapping Practice & Cartographic Representation*. Chichester: John Wiley & Sons, pp. 89–101.

Cosgrove, D. & Domash, M. (1993) 'Author and authority: Writing the new cultural geography' in Duncan, J. & Ley, D. (eds) *Place/Culture/Representation*. Abingdon: Routledge, pp. 25–28.

Cresswell, T. (2013) *Geographic Thought: A Critical Introduction*. Chichester: John Wiley & Sons.

Crouch, D. (2010) 'Flirting with space: Thinking landscape relationally', *cultural geographies*. 17(1), pp. 5–18.

Daniels, S. & Cosgrove, D. (1988) 'Introduction: Iconography and landscape' in Daniels, S. & Cosgrove, D. (eds) *The Iconography of Landscape: Essays on the Symbolic Representation, Design and Use of Past Environments*. Cambridge: Cambridge University Press, pp. 1–10.

De Certeau, M. (1984) *The Practice of Everyday Life*. Berkeley: University of California Press.

Del Casino, V. J. & Hanna, S. P. (2006) 'Beyond the "binaries": A methodological intervention for interrogating maps as representational practices', *ACME: AN International Journal for Critical Geographies*. 4(1), pp. 34–56.

Deleuze, G. & Guattari, F. (1987) *A Thousand Plateaus: Capitalism and Schizophrenia*. London: Athlone Press.

DeLysier, D. & Sui, D. (2014), 'Crossing the qualitative quantitative chasm III: Enduring methods, open geography, participatory research, and the fourth paradigm', *Progress in Human Geography*. 38(2), pp. 294–307.

Derrida, J. (1976) *Of Grammatology*. Baltimore: John Hopkins University Press.

Dirksmeier, P. & Helbrecht, I. (2008) 'Time, non-representational theory and the "performative turn"—towards a new methodology in qualitative social research', *Forum: Qualitative Social Research*. 9(2), Article 55.

Dixon, D. P. & Jones III, J. P. (1998) 'My dinner with Derrida, or spatial analysis and post-structuralism do lunch', *Environment and Planning A*. 30(2), pp. 247–260.

Duncan, J. & Duncan, N. (1992) 'Ideology and bliss: Roland Barthes and the secret histories of landscape' in Barnes, T. & Duncan, J. (eds) *Writing Worlds: Discourse, Texts, and Metaphors in the Representation of Landscape*. London: Routledge, pp. 18–37.

Gallagher, M. and Prior, J. (2014) 'Sonic geographies: Exploring phonographic methods', *Progress in Human Geography*. 38(2), pp. 267–284.

Garrett, B. (2011) 'Videographic geographies: Using digital video for geographic research', *Progress in Human Geography*. 35(4), pp. 521–541.

Haraway, D. (1988) 'Situated knowledges: The science question in feminism and the privilege of partial perspective', *Feminist Studies*. 14(3), pp. 575–599.

Harley, J. B. (1990) 'Cartography, ethics and social theory', *Cartographica*. 27(2), pp. 1–23.

Harley, J. B. (1989) 'Deconstructing the map', *Cartographica*. 26(2), pp. 1–20.

Harley, J. B. (1987) 'The Map as Biography: Thoughts on Ordnance Survey Map, Six-Inch Sheet Devonshire CIX, SE, Newton Abbot' in Dodge, M., Kitchen, R. & Perkins, C. (eds) (2011) *The Map Reader: Theories of Mapping Practice & Cartographic Representation*. Chichester: John Wiley & Sons, pp. 327–331.

Harmon, K. A. (2004) *You Are Here: Personal Geographies and Other Maps of the Imagination*. New York: Princeton Architectural Press.

Hawkins, H. (2017) *Creativity*. Abingdon: Routledge.

Highmore, B. (2002) *Everyday Life and Cultural Theory: An Introduction*. London: Routledge.

Holloway, L. & Hubbard, P. (2001) *People and Place: The Extraordinary Geographies of Everyday Life*. Harlow: Pearson Education Ltd.

Holtzman, J. D. (2006) 'Food and memory', *Annual Review of Anthropology*. 35, pp. 361–78.

Ingold, T. (2000) *The Perception of the Environment: Essays in Livelihood: Dwelling and Skill*. London: Routledge.

Kitchin, R. & Dodge, M. (2007) 'Rethinking Maps', *Progress in Human Geography*. 31(3), pp. 331–344.

Kitchin, R., Gleeson, J. & Dodge, M. (2012) 'Unfolding mapping practices: a new epistemology for cartography', *Transactions of the Institute of British Geographers*. 38(3), pp. 1–17.

Kitchin, R., Perkins, C. & Dodge, M. (2009) 'Thinking about maps' in Dodge, M., Kitchin, R. & Perkins, C. (eds) *Rethinking Maps*. Abingdon: Routledge, pp. 1–25.

Krygier, J. B. (1995) 'Cartography as an art and a science?', *The Cartographic Journal*. 32(1), pp. 3–10.

Latham, A. (2003) 'Research, performance and doing human geography: Some reflections on the diary-photograph, diary-interview method' in Price, P. L. & Oakes, T. S. (eds) (2008) *The Cultural Geography Reader*. Abingdon: Routledge, pp, 68–76.

Latham, A. & McCormack, D. P. (2004) 'Moving cities: Rethinking the materialities of urban geographies', *Progress in Human Geography*. 28(6), pp. 701–724.

Laurier, E. & Philo, C. (2006) 'Possible geographies: A passing encounter in a café', *Area*. 38(4), pp. 353–363.

Leach, N. (2002) 'Introduction' in Leach, N. (ed.) *The Hieroglyphics of Space: Reading and Experiencing the Modern Metropolis*. London: Routledge, pp. 1–12.

Lorimer, H. (2005) 'Cultural geography: The busyness of being "more than representational"', *Progress in Human Geography*. 29(1), pp. 83–94.

Lorimer, J. (2010) 'Moving image methodologies for more-than-human geographies', *cultural geographies*. 17(2), pp. 237–258.

Marks, L. U. (2000) *The Skin of the Film: Intercultural Cinema, Embodiment, and the Senses*. Durham: Duke University Press.

Massey, D. (1997) 'Spatial disruptions' in Golding, S. (ed.) *The Eight Technologies of Otherness*. London: Routledge, pp. 218–225.

Merleau-Ponty, M. (2002) *Phenomenology of Perception*. 1st edn 1962. London: Routledge.

Nash, C. (2000) 'Performativity in practice: Some recent work in cultural geography', *Progress in Human Geography*. 24(4), pp. 653–664.

Nold, C. (2007) Stockport Emotion Map. Available at: http://stockport.emotionmap.net (Accessed: 29 December 2017).

Nold, C. (2004) *Bio Mapping/Emotion Mapping*. Available at: www.biomapping.net (Accessed: 29 December 2017).

Pickles, J. (2004) *A History of Spaces: Cartographic Reason, Mapping and the Geo-coded World*. London: Routledge.

Pink, S. (2015) *Doing Sensory Ethnography*. London: Sage.

Proust, M. (2013) *In Search of Lost Time, Volume 1: Swann's Way*. New Haven: Yale University Press.

Seremetakis, N. (1994) 'The memory of the senses, part 1: Marks of the transitory', in Seremetakis, N. (ed.) *The Senses Still: Perception and Memory as Material Culture in Modernity*. University of Chicago Press: Chicago, pp. 1–18.

Shannon, E. E. & Weaver, W. (1949) *The Mathematical Theory of Communication*. Chicago: University of Illinois Press

Shields, R. (1991) *Places on the Margin: Alternative Geographies of Modernity*. London: Routledge.

Sutton, D. (2001) *Remembrance of Repasts: An Anthropology of Food and Memory*. London: Bloomsbury.

Thrift, N. (2008) *Non-Representational Theory: Space\Politics\Affect*. Abingdon: Routledge.

Thrift, N. (2000) 'Dead or alive?', in Cook, I., Crouch, D., Naylor, S., & Ryan, J. (eds) *Cultural Turns/Geographical Turns: Perspectives on Cultural Geography*. Harlow: Prentice-Hall, pp. 1–6.

Thrift, N. (1997) 'The still point', in Pile, S. & Keith, M. (eds) *Geographies of Resistance*. London: Routledge, pp. 124–151.

Travers, M. (2009) 'New methods, old problems: A skeptical view of innovation in qualitative research', *Qualitative Research*. 9(2), pp. 161–179.

Vannini, P. (2015) 'Non-representational research methodologies: An introduction' in Vannini, P. (ed.) *Non-Representational Methodologies: Re-envisioning Research*. Abingdon: Routledge, pp. 1–18.

Wood, D. (2002) 'The map as a kind of talk: Brian Harley and the confabulation of the inner and outer voice', *Visual Communication*. 1(2), pp. 139–161.

Wood, D. (1993) 'The fine line between maps and mapmaking, *Cartographica*. 30(4), pp. 50–60.

Wood, D. (1992) 'How maps work' *Cartographica*. 29(3&4), pp. 66–74.

Wood, D. & Fels, J. (2008) *The Nature of Maps: Cartographic Constructions of the Natural World*. Chicago: University of Chicago Press.

3   Art-geography collaboration and the potential of graphic design

Introduction

As we have seen previously, there has been a 'spatial turn' within the arts, and artists have been 'consciously and critically deploying the skill sets of other disciplines'; for example, the artist as ethnographer or as cartographer (Hawkins 2012: 64). Traditional research methods such as mapping and fieldwork have become popular, with fieldwork being undertaken in a myriad of ways and contexts, for example through the practice of walking, dance and hot air ballooning (Daniels, Pearson & Roms 2010: 2). The discipline of geography has been described as going through a 'creative (re)turn' that has evolved from the study of creative products such as art works, novels, music and films in cultural geography, to the adoption of creative practices such as writing, curating, film-making and art in all its forms, which offer 'a particular methodological value' for geographers (Hawkins 2017: 11). This shift to the incorporation of creative methods is largely due to debates around issues such as non-representational theory, performativity and phenomenology, along with shifts within qualitative social science research methods in general (see for example, O'Neill 2008; O'Neill & Hubbard 2010; Roberts 2008) which, rather than attempting to reveal certainty or truth, adopt an embodied, open and reflexive stance in relation to interpreting the complexity of the world. Thus, such research methods and 'interpretive strategies' enable geographers and others to 'capture the ephemeral, the fleeting [and] the immanence of place (Davies & Dwyer 2007: 261).

According to Davies & Dwyer (2010: 91), the 'dialogue between cultural geography and artistic practice has a long history' and even though constraints within traditional academic publishing and the demands of university research assessment exercises the world over might have been expected to constrict some of this experimentation, formal opportunities for collaboration are increasing. Indeed, the rise in geographer-artist collaborations over the past two decades has been described as 'exponential' with such collaborations almost representing 'a new orthodoxy within the discipline' (Tolia-Kelly 2012: 135). However, Tolia-Kelly goes on to suggest that such work is best framed as visual culture rather than art, and views a practice-led approach that is 'informed by design, pattern and impression' as one where geographers might more productively position their 'visual research edge' (Tolia-Kelly 2012: 139). This reference to design is extremely unusual within the literature on creative methods and this book goes some way towards beginning to redress that balance and open up a space for productive dialogue. The latter part of the chapter therefore explores the potential graphic design and typography in particular have in creating a re/presentation and experience of place that echoes contemporary definitions of it as complex, ongoing and relational.

Art-geography collaboration and creative methods

DeSilvey & Yusoff (2006: 753) state that 'art and geography share a common route in the search for knowledge through the medium of vision' and that 'art plays an important role in questioning geography's visual methods' (DeSilvey & Yusoff 2006: 574). They note that the works of contemporary artists have increasingly begun to reference geographical forms such as maps, field notes and charts. However, geography and art do not necessarily share 'a common language or set of practices' and are perhaps best described as discrete disciplines that intersect at the point of their shared interest in the visual (DeSilvey & Yusoff 2006: 582) – the spatial should also be added to this. Mirroring Thrift's claims discussed in the previous chapter, Ryan (2003) suggests that many of the methods used by cultural geographers are conservative and could be more imaginative, and he proposes dialogue and collaboration with visual artists as a route to new ideas.

Opportunities for dialogue

Spaces for such dialogue are now increasingly available and one of the first of these was established in 2000 with the addition of the section *Cultural Geographies in Practice* in the journal *cultural geographies*. The section not only provides a forum for those cultural geographers using alternative methods within their research, but also for non-geographers working with cultural geographic themes and ideas. It also offers a space for work that doesn't conform to the traditional format of a text-based article. Since then a raft of special calls, conferences, conference strands, journals and research centres focusing on this growing area of interest have been established. For example, the University College London *Urban Lab* was established in 2005 to bring together scholars and practitioners from across the arts and sciences in order to experiment with and develop new methods of research in an urban context that eschew traditional disciplinary boundaries and actively seek cross- and interdisciplinary collaboration. The biennial *International Visual Methods Conference* was established in 2009, with the theme of the 2017 conference being 'Visualising the City'. In 2013, *Geographical Review* published a special issue entitled 'Geography and Creativity' featuring contributions from 'practicing artists/writers, academic geographers, and those who blur these discrete groupings' (Marston & De Leeuw 2013: iv). The *GeoHumanities* journal was launched in 2015 and focuses on methodological and conceptual debates within geography and the humanities; encourages work that is being done at the intersection of geography and a range of disciplines within the humanities; and provides a forum for critical reflections on artistic work that has a geographic context. Much like *cultural geographies, GeoHumanities* is also formed of two discrete sections, one which features traditional, full length scholarly articles, the other – the *Practices and Curations* section – featuring shorter pieces that blur the lines between creative practice and the academy. In 2016, The UK Arts and Humanities Research Council (AHRC) and Engineering and Physical Sciences Research Council (EPSRC) launched a joint call for an immersive experience project which sought to link researchers from both the social sciences and technology with those from the arts and humanities, along with creative economy/industry partners – the three key themes of the call were memory, place and performance. Also in 2016, the *Centre for GeoHumanities* was established at Royal Holloway, University of London, linking geographers, arts and humanities scholars and the heritage, cultural and creative sectors. Its aim is to showcase and nurture work within the arts and humanities that has a strong geographical connection, for example work that centres on space, place, landscape or environment.

Creative collaborations

This burgeoning number of examples would seem to indicate that creative research projects between geographers and artists have the potential to enable a proactive sharing of ideas and the development of new visual methodologies. In any cross- or interdisciplinary work, tensions may well arise due to the collaborators' differences. Whilst such tensions can be productive, challenging 'ideas of authority, expertise, and established ways of working' (Nash 2013: 51), there are also instances where both artist and academic geographer may feel they have to defend their position and the result is simply for the divide to remain uncrossed (Nash 2013: 52).

One such art-geography collaboration, *Visualising Geography*, was an AHRC funded research project developed by geographers Felix Driver and Catherine Nash, and artist Kathy Prendergast. Based in the department of geography at Royal Holloway, University of London, the project sought to develop long-standing connections between geography and the visual arts. *Landing*, the work in progress exhibition that developed from the research project was curated by Ingrid Swenson and staged at Royal Holloway during summer 2002. *Visualising Geography* actively sought to avoid the simplistic binary oppositions such as 'academic' and 'creative' and saw all participants as equal partners in the collaborative process. The artists, therefore, were not simply there to illustrate the geographers' ideas (Driver et al. 2002). The workshop discussions, held as part of the project, highlighted the potential for reframing the terms artist and geographer. Differences in approach were revealed with artists positioning and redefining ethnography as art, and academics drawn to creative research methods encountering artists whose practice was based on rigorous empirical methods. From these conversations disciplinary boundaries 'began to shift, realign, dissolve and sometimes re-crystalise' (Driver et al. 2002: 8). For some, these conversations led to productive collaborations, however for others, the differences were impossible to surmount and ultimately destructive, resulting in 'senses of diffidence, insecurity, defensiveness, resistance or assumptions of authority' (Driver, Nash & Prendergast 2002: 9). Tolia-Kelly (2012: 137) suggests that many such collaborations result in 'each researcher modestly sticking with their expertise' and indeed many of the collaborations in *Landing* didn't result in 'the actual co-production of the art itself' (Nash 2013: 51). Nash also suggests that it is perhaps easier to 'relinquish a degree of academic authority than to achieve acceptance as a creative practitioner' given the particular skillset involved and years of training that artists have usually undertaken (Nash 2013: 53). However, both Tolia-Kelly (2012: 137) and Jellis (2015: 369) suggest that the reverse may also be an issue, as rarely is there any discussion of the benefits of the artist immersing themselves in a geographic understanding of everyday life and place. Interestingly, Schaaf, Worrall-Hood & Jones (2017) discuss an art-geography collaboration undertaken with students from both disciplines and speculate as to whether many of the disciplinary traits displayed that either disrupted or developed the collaborative relationship are the result of pedagogical differences reinforced at an earlier stage of education.

Another example of collaboration is *Domain* (Kinman & Williams 2007), a project between cartographer Edward Kinman and artist John Williams. Their project centred on the representation of Longwood University campus in Virginia and the visual outcome was a series of four stoneware panels. The panels depicted the campus and its adjacent ethnic neighbourhood from diverse social, cultural and political historical perspectives. Williams' focus and contribution to the partnership seems to centre on the media involved, enabling Kinman to execute his ideas through clay. Whilst this outcome may be related to Nash's point above, as it is unlikely that Kinman would have felt comfortable working in clay without any prior knowledge or training, it does seem as if the collaboration failed to challenge both roles. So, although Kinman reflects on how working with an alien media changed his cartographic process, this collaboration seems to delimit, or perhaps presume, the role or 'domain' of the artist. A more reflective and productive experience seems to be that of artists Annie Lovejoy and geographer Harriet Hawkins (2009) whose collaboration resulted in *Insites*, an artists' book. They endeavoured not to avoid their disciplinary differences, but to find a way they could work together. Artist Kate Foster and geographer Hayden Lorimer (2007: 427–428) also used collaboration 'to ask awkward questions of your own conventions and accepted working practice', and realised that it is possible to enmesh aspects of joint-working into their individual work, and that at points it was possible to move between each other's perceived disciplinary boundaries and exchange both roles and skills.

Creative methods

However, as Nash (2013) concedes, there may be a danger, inherent in the pursuance of creative methods, that a focus on aesthetics, form and new technical possibilities takes precedence over research content, and this is clearly something to be avoided. So whilst there are productive possibilities in both collaboration and creative approaches, these should not be seen as superseding traditional research methods or conventional formats. Indeed, many geographers are reframing traditional methods within the context of a variety of creative approaches. For example, Lorimer & MacDonald's (2002) use of 'rescue archaeology' on Taransay attempts 'to rework traditional approaches' in order to devise a field methodology that is appropriate for undertaking 'an archaeology of the present' (Lorimer & MacDonald 2002: 95); and Wylie's (2006) *Smoothlands: fragments/landscapes/fragment* combines walking and 'phenomenological methods of watching and picturing' (Wylie 2006: 458). Rather than framing these experiences and images within a traditional written account from the field, Wylie interweaves his photographic images of place with short excerpts of writing about landscape and subjectivity, read whilst undertaking his journey. Pryke's (2002) study of Potsdamer Platz in Berlin 'employs forms of visual and audio montage to show something of the making of spaces' (Pryke 2002: 474). The methods discussed all inextricably link to traditional geographic methods of exploration and fieldwork, but are being reframed and reworked in new ways. More recently, DeSilvey's (2012) work on Mullion harbour in Cornwall was presented in *cultural geographies* as 'a reverse chronology of (present tense) moments gleaned from diverse sources ranging over three centuries' (DeSilvey 2012: 36). This approach is supplemented by the addition of contemporary auto-ethnographic anecdotes and paired, asynchronous photographic records as a form of visual montage. The piece draws together diverse strands within the narrative which are 'designed to reveal that the harbour's apparent stability hides a precarious history, and to draw out patterns and themes that are obscured by more conventional tellings' (DeSilvey 2012: 48). Coles' (2014) recent work on Borough Market utilises both image and text in what he describes as 'kind of visual "topography"' (2014: 515). As we have seen in discussions of traditional mapping practices in Chapter 2, a topographical approach has its roots in a positivist scientific history of 'imperial and geographical subjugation' (Coles 2014: 516). However, in this work, Coles' positions topography as a 'methodological encounter with place' that uses a range of theoretical and empirical texts (here he is using the concept of text in its widest form) to 'engage with place and place-making and to (re)present the interrelations between its "surfaces" and "structures"' (Coles 2014: 515). His topography uses many of the empirical methods usually undertaken as part of ethnographic work, and whilst Coles positions topography as a methodological approach like ethnography, he suggests it differs by being a 'spatial practice where place emerges through its writing, telling and its (re)presentation' (Coles 2014: 516).

Creative methods have also been used by many geographers in a participatory context. The use of such methods in this context offers two clear advantages. First, it engages the participants in ways that could be described as less formal or pressurised than many traditional research methods such as surveys or interviews. Second, it engages the participants and researcher in a form of collaboration, giving the participants a sense of control and ownership over their contribution. For example, Parr (2007) has used collaborative film making as both a research process and a practice with people who have severe mental health problems. She found that the process of editing acted similarly to the process of traditional qualitative analysis and therefore enabled discussions with collaborators which clarified or established further meanings and understandings of living with a severe mental health condition. The idea of analysis being undertaken via the process of creative practice will be discussed further in the following chapter.

As well as reinterpreting traditional fieldwork methods, geographers are also engaging with new technologies – film and sound in particular – as ways of exploring and representing place. Lorimer's (2010) paper *Moving image methodologies for more-than-human geographies* addresses the growing interest in moving image 'for grasping the . . . non-representational dimensions of life' (Lorimer 2010: 237) and 'contributes to ongoing efforts to develop practical visual methodologies for cultural geography' (Lorimer 2010: 251). Gallagher & Prior's (2014) paper *Sonic geographies: Exploring phonographic methods* highlights the potential for sound to be used as a research method in ways that go beyond the more traditional uses of audio such as recording interviews. Cultural geography is obviously richer for these visual and aural developments, but many of these newer technologies, although partly chosen due to debates of a non-representational nature, also seem to be a conscious rejection of the old. There is a sense that print equates to the immovable, the non-relational, the closed rather than the open, that text should be relegated to the tired, dusty pages of academic journals that refuse to move with the times. Even within the pages of *Cultural Geographies in Practice* few of the articles featured really explore the communicative possibilities of type, image and the page.

Writing: a mode of telling

The potential offered by text is of concern to some geographers and in their study of creative writing, Brace & Johns-Putra (2010), suggest that the practice of crafting a text is performative. By taking this position they address the divide between thought and action referred to by Nash (2000). They glimpse: 'a fusion of thought, action, body and text in ways that undermine the epistemological separation of representation and non-representation and thereby avoid the critique of representation that emphasises its static fixity and evacuation of process' (Brace & Johns-Putra 2010: 403). Similarly, Lorimer (2008: 182) seeks to further develop cultural geography's relationship with the word, suggesting that: 'What a geographical education does not always equip us with is a way with words; a language sufficient to do fullest justice to the intensities, to the properties and to the rich lore of place'. In these pieces, both Lorimer and Brace & Johns-Putra focus on creative writing and the potential of a poetic engagement with place. Lorimer notes that there 'seems an increased premium . . . placed on the creative performance, presentation and writing of geographic studies of place' and goes on to suggest that geographers are now prepared to 'consider style as a pressing issue' within their writing (2008: 182). This seems to assume that traditional academic geographic writing is devoid of style. However, as Kinross (1985) has pointed out in relation to Modernist typography, and Atkinson (1990: 2) in relation to ethnography, this is merely a 'rhetoric of neutrality' and such writing clearly adopts the style of the academy. Perhaps part of the issue is that the tone or style is so overtly present within most academic geographic writing that it ceases to be seen as a choice, but is rather a natural process that happens despite the author not because of her. In a similar vein to Lorimer, Springer (2017) argues for an embracing of '*geopoetics*', an approach to earth writing that 'releases our geographical imaginations from the shackles of our disciplinary past' (Springer 2017: 2; italics in original). Although, unlike Lorimer, Springer's view is that the perpetuation of what might be described as the academic style is prevalent within the academic geographic publishing world, with peer reviewers intent on disciplining those who do not follow the 'correct' way to do geography (Springer 2017: 5). Springer's intention is for geopoetics to disrupt the writing landscape, to challenge the perception that there is one right way to disseminate research or construct a narrative around a sense of everyday life and place. As he points out, since the crisis of representation any kind of writing 'can no longer be considered a straightforward process of conveying the real world' (Springer 2017: 7). Thus, taken to its logical conclusion, there can be no single 'correct' form of geographical writing, for any writing about place is only ever a partial view from one perspective. Thus, Springer's concern is less to replace traditional forms of academic writing with a metaphorically driven geopoetics, and rather to advocate for broader acceptance of work that spans a range of epistemological, ontological and methodological approaches (Springer 2017: 12).

Reading Springer's article there is a sense that he sees this poetic approach as one of a leftfield minority, and though traditional academic publishing is still clearly in the majority, several other geographers are also exploring the potential of creative writing as part of their research and practice. Cresswell notes that such creative writing practices are 'informed by phenomenology, poststructuralism, and assemblage theory (among other approaches)' and 'represent a practice of 'place-writing' that attempts to grapple with particular places in all their complexity' (Cresswell 2015: 56–57). The strategies used in such works are often the development of a text that is collage- or montage-like, evoking Benjamin's (1999) approach to place in *The Arcades Project* and aligning with Massey's ideas of place in process (1994, 2005) in the presentation of a narrative that adopts a variety of voices and perspectives. A more contemporary example is Price's (2004) *Dry Place: Landscapes of Belonging and Exclusion* – 'an exploration of how we relate (to) place' centred on a study of the border area between Mexico and the USA (Price 2004: viii). The book weaves together a range of narratives and includes some of her own poetry in the closing chapter. In referring to the idea of landscape as text, Price suggests she wishes to take this further in the writing of *Dry Place*:

> Places, as well as the landscapes that allow us to grasp them, are thoroughly narrative concepts. They would not exist *as places* were it not for the stories told about and through them. Stories constitute performative, mimetic acts

that conjure places into being and sustain them as the incredibly complex, fraught constructs that they are. Stories—particularly those tales we tell our-selves about ourselves—provide the weathered bedrock that binds human collectives to the lands we inhabit.

(Price 2004: xxi; italics in original)

Price essentially views place as a narrative construct and the stories that make place are constitutive of our identities. Yet this narrative, and thus place, is poly-vocal in nature, 'riddled with silences, dissenting voices, and differences as it is with unisonant consensus' (Price 2004: xxii). The power of narrative, is therefore central to the creation, understanding and representation of place.

Narrative and storytelling

In an academic context, the majority of research is disseminated in written form. In their search for methods that might enable writing that communicates a more evocative sense of place to the reader, cultural geographers and others have looked towards creative writing and poetry, and underpinning those methods is a more fundamental approach to constructing a text, that of narrative. Narrative is perhaps originally associated with fiction writing, yet given that we now under-stand all texts to be partial truths, storytelling is resurgent in cultural geography and across the social sciences. Equally, stories, and the telling of them, are criti-cal to the understanding and representation of everyday life and place. Indeed, Massey suggests that 'If space is . . . a simultaneity of stories-so-far (rather than a 'surface'), then places are collections of those stories, articulations of the wider power-geometries of space' (Massey 2005: 130).

After the 'crisis of representation', stories and storytelling inevitably fell out of favour with geographers as they were seen to contribute to continued power imbalances within society. However, more recently, the potential of storytelling to contribute to non-representational approaches to place has seen an increase in the use of story to create experiential and imaginative representations in both verbal and non-verbal formats (Daniels & Lorimer 2012: 6). Stories and storytell-ing are being used in different ways; stories as objects of knowledge; storytelling as a practice or form of academic expression; as creative or fictional expressions; as vehicles for the personal experience of the world of research participants; and as autoethnographic stories (Cameron 2012: 575–576). Within this range of approaches, Cameron (2012) suggests three key areas in which the majority of contemporary geographic stories and storytelling fall. First, as 'small stories' (Lorimer 2003) that express personal experience and 'lives in all their particu-larity and mundanity' (Cameron 2012: 575). In this contemporary context, such stories aren't being framed within concepts of power and ideology or as in opposi-tion to stories of a larger or global scale. Rather, they are positioned as a proactive turn towards 'thicker descriptions and understandings of the small' (Cameron 2012: 577). Second, as a way of challenging and transforming existing power relations, with individual or marginalised stories used as a counterpoint to the metanarratives of place (Cameron 2012: 575). In particular, Gibson-Graham sees the performative potential in narratives to create change in the economic world and to function as 'ontological interventions' (2008: 614) So rather than simply explain what is happening in the world, academic writing and theory can be used 'to enact a revolution of sorts' (Gibson-Graham 2014: S151). In this context it is the story itself, the practice of storytelling, and the 'capacity for stories to be practiced in place' (Cameron 2012: 581) that challenge the economic status quo. Third, many geographers are using narrative to convey personal, experiential geographies in the context of autoethnographic, non-representational work. In this context, story is being used as an 'affective tool' with which to engage audiences and move them 'towards new realms and practice' (Cameron 2012: 575). Thus, the conception of stories and storytelling shifts beyond that of a solely representational form and practice, to one in which they have the capacity to 'move, emerge, and affect in the very act of their telling' (Cameron 2012: 588).

Poetic representations of place

Examples of geographers using poetry, or poetry being used in a geographical context are also prevalent. For example, in the journal *Social and Cultural Geography*, Eshun & Madge (2016) discuss the way poetry 'might provide further fresh insights for the creative (re)turn in geography' and, in particular, 'whether it might enable creative geographies [to] become more attentive to a pluriversal world perspective' (Eshun & Madge 2016: 778). A special issue of *Geographical Review* on geography and creativity featured poems from novelists and poets Laisha Rosnau (2013: 139–142) and Gillian Wigmore (2013: 256–259) and geographer Tim Cresswell (2013: 285–287). Geographer and writer Sarah de Leeuw (2012; 2015) has also published two volumes of poetry.

Whilst poetry is written, it is also spoken, and Lorimer takes his argument about creative writing further and suggests that it can benefit from being read aloud. 'Simply put, language transforms when it is heard, rather than read on the page, such that listeners might reasonably claim to see the sounds spoken' (2008: 182). Online publishing is enabling geographers to explore such strategies and in 2011 *ACME: An International Journal for Critical Geographies* published – '*The Bus Hub*' – a poem/song written and composed by Kafui Attoh a master's student in geography at Syracuse University (2011: 280–285). The written poem was accompanied by a link to the song that the reader/listener could download. However, it wasn't simply the audio format that challenged the journal's editor, but the poem also – even though the journal seeks to publish work in alternative formats. Editor David Butz was concerned that the poem did not fulfil 'the journal's editorial mandate: to publish "critical and radical analyses of the social, the spatial and the political"'. The word that troubled him was 'analyses' and he asked, 'What place does a submission that operates evocatively rather than analytically have in a journal dedicated to analysis?' (Butz 2011: 278). Ultimately, the piece went out to three reviewers using the standard criteria, but with the addition of a supplementary question as to whether the piece should be published alone or with an additional analytical discussion (Butz 2011: 278). Ultimately '*The Bus Hub*' was published with a short introduction from the author and followed by two texts from academics who were asked to reflect on the 'epistemological implications of understanding an evocative creative piece like this as geographical scholarship or representation' and/or 'their experience of it as geography' (Butz 2011: 279). As with any bus hub the world over, the bus hub in downtown Syracuse is constantly shifting and changing, with people passing through and waiting, the noise of engines, the sound of chatter, the smell of petrol, oil, exhaust fumes, cigarettes and fast food. Bus hubs are also places where those who have to rely on public transport congregate and those who need warmth also often find a space to wait.

> They meet at the intersection of urban decay and the carnivalesque
> It is a walk-able street but people are
> Standing in place
> Black folks on the corner
> Poor people and their problems
> Out in the open, like an exposed wound
> The upper echelon wants them hidden and gone
> Buried under the rug.
>
> (Attoh 2011: 283)

A similar description of such a scene is perhaps easy to imagine in a more traditional piece of ethnographic writing, but whether that version would have either the courage or the imagination to use phrases like 'an exposed wound' is debatable. This is perhaps an example of Springer's desire for a geopoetics, and de Leeuw in her commentary suggests that Attoh's poetry and song are 'balanced on the borderlands' and are as much about 'claiming new spaces and (re)making places on the margins' of geography, as they are a 'work about people and places that are doing those same-said activities' (De Leeuw 2011: 291).

Lorimer and others exploring the potential of sound in the context of creative writing are unlikely to be trying to privilege the spoken word over the written in the way that many philosophers have done for centuries. However, this is perhaps symptomatic of a drive to explore new areas within cultural geography – in this case, creative writing and performance – that comes at a cost of ignoring the potential of language in print. For example, within sound and poetry in particular there is a rich tradition of typographic experimentation from Dada, to the Futurists, and to concrete poetry. For example, in discussing the work of French poet Stephane Mallarmé – whose work explored the potential of content and form – Eco notes the following:

> Blank space surrounding a word, typographical adjustments, and spatial composition in the page setting of the poetic text—all contribute to create a halo of indefiniteness and to make the text pregnant with infinite suggestive possibilities.

> The search for *suggestiveness* is a deliberate move to 'open' the work
> to the free response of the addressee. An artistic work that suggests is also
> one that can be performed with the full emotional and imaginative resources
> of the interpreter.
>
> (1989: 9; italics in original)

Yet the majority of academic research outputs are still through the traditional text
based journal article, even if the methods used in the research have been creative
in some way (see Davies & Dwyer 2010: 92; Lorimer 2005: 89; Thrift 2000: 3).
However, typography and graphic design can offer much more than a range of
typefaces or fonts to choose from and the construction of a page in an academic
journal that enables the reader to digest the ideas contained without visual inter-
ference. Therefore, as Nash also suggests, there is the potential to work with the
formats in order to 'harness the expressive potential of writing and speaking to
produce evocative, compelling, and convincing narrative accounts of their subjects' (Nash 2013: 53–54). Not only is there potential but, given that 'impact' and
forms of public engagement are increasingly part of the requirements linked to the
government research funding of universities, it is very likely that this turn to visual and creative methods will be sustained, and perhaps increase (see Tolia-Kelly
2012: 137; Nash 2013: 54). Tolia-Kelly sees the potential in design, suggesting
that rather than position such collaborations and methodological inventiveness
within the context of art, they are better located within visual culture – identifying
that visual culture centres on a practice that is informed by design, pattern and
impression, as opposed to a 'self-consciously lived, artistic expression' (Tolia
Kelly 2012: 139). This reference to design is extremely unusual within the literature discussing creative methods, and we therefore now turn our focus to a discussion of graphic design and typography. In engaging with both form and content,
and eschewing a disciplinary focus within graphic design that has previously been
on aesthetics and notions of 'style', design process and practice have much to
offer geographers seeking to develop visual research methods. Even within the
confines of print-based work, graphic design and typography offer the potential to
develop understandings and re/presentations of place that engage the reader in an
embodied, multi-sensory experience.

Graphic design: from form to content

The function of graphic design and typography, in basic terms, is to communicate
by giving visual form to content. The variety of this content is enormous – from
signage, to fashion advertising, to pharmaceutical packaging, and everything else
in between and beyond. The graphic designer uses typography, colour, image
and format to produce a visual solution that meets the requirements of the cli-
ent's desired message. Consideration of the audience is part of the process, and
according to Buchanan the goal is 'to induce in the audience some belief about
the past . . . the present . . . or the future' (1989: 92). The use of rhetoric enables
the audience to become a 'dynamic participant' as the designer persuades through

argument rather than statement (Tyler 1992: 22). Although the audience is active in this process, this position is rooted in the belief that one meaning can be communicated – that of the author, via the designer. One could perhaps argue that having a very narrow target audience with a shared set of cultural values and experiences, and a simple message, this might be possible. However, the representation of place is a complex proposition.

Frascara (2006: xiv) states that the designer should create a 'space' where people engage with a message and develop interpretations. This acknowledges that there is a possibility of multiple meanings being developed by a variety of readers and the idea of a 'space of interpretation' provides an interesting model to pursue with regard to place (see Barnes 2012). This 'space' is not only created through language – all elements are involved – from paper stock and format, to typeface and colour choice. This could be described as 'visual language', or perhaps style.

The surface of the page

In his book *Lines: A Brief History* (2007), anthropologist Tim Ingold offers an analogy between the map and the printed page that draws the disciplines of graphic design and geography together and articulates an issue central to this book. Ingold (2007: 24) states that the page – like the map – has 'lost its voice'. In the same way that there are no traces of the life of a place on a map, no evidence of the journeys undertaken to gain the knowledge to create it, language is also 'silenced' on the page of a book through the mechanical process of print. Ingold suggests that before the advent of print there was a clear link between the manual gesture of writing and its graphic form on the page, yet now this link is broken. With this link severed, Ingold (2007: 26) states that, in terms of the page, the perception of 'surface' has shifted from: 'Something akin to a landscape that one moves through, to something more like a screen that one looks at, and upon which are projected images. . .' In geographic terms however, landscape is often positioned as something the viewer remains outside of – it is 'an intensely visual idea'. In contrast, places are 'very much things to be inside of' (Cresswell 2015: 17). So perhaps Ingold's statement should read as 'something akin to *place* that one moves *through*, to something more like *landscape* that one looks *at*. . .'

Ingold posits that prior to the invention of printing technology, writing was akin to drawing, and that words penned by the scribes conveyed feeling in an expressive way that is no longer apparent in print (2007: 3). Also, in medieval times, reading was not solely related to cognition, there was a performative element to it as most texts, even if read in private, were read aloud. Therefore, it was both an 'acting out' and 'a taking in' (Ingold 2007: 17) so listeners were, in effect, 'using their eyes to hear' (Ingold 2007: 13). However, with reading becoming an internal, silent activity, and print negating the unmediated connections 'from the prophet's mouth to the scribe's inky traces' (Ingold 2007: 13), Ingold suggests that the reader became more passive and began to use their 'ears to look' (Ingold 2007: 13) – hence Ingold's use of the screen as a metaphor for the page. This 'surveying' of the page (Ingold 2007: 92) also brings to mind notions of the map positioning the viewer above and outside the territory. In relation to print, this view of reading is one that separates thought and action. However, the human link with the printed word remains in the form of the graphic designer who has the ability to recreate the page as 'place' through an expressive use of typography. In doing so, this also reconfigures the act of reading as both cognitively and perform-atively embodied. However, Ingold (2007: 26) not only sees print as silencing the page, but as introducing a 'split of skilled handicraft into separate components of "imaginative design" . . . and "merely" technical execution', that leads to 'the implementation of pre-determined operational sequences that could just as well be done by machine'. Given that scribes executed work written or narrated by others, it is hard to see how this is actually very different, and with the advent of the *Apple Mac* and desktop publishing, designers actually now have more control over the 'technical execution' than they did previously. Perhaps if one reflects on the process of constructing a text in *Microsoft Word* it is understandable how Ingold might have arrived at this statement as it often pre-empts decisions regard-ing the formatting of typography.

Ingold's thinking seems to stem from a belief that the notion of 'craft' – implying a physical engagement with tools and materials – no longer exists due to the advent of technology. Generally, however, proactive choices are made before one presses 'command P', and graphic design education urges students to move beyond passively accepting the default choices on design software, and to view the computer as a tool. Graphic designers also continue to engage with the physi-cality of media through the selection of stock, format and binding, and perhaps even die-cutting or embossing if budget allows. They also physically prototype their work prior to final production. I would suggest that this also goes some way to challenging Ingold's perception of separate components of design and execu-tion. Pages may no longer reveal the visible nuances of the scribe's hand, but they are no less a physical space upon which the mark of the designer is clearly present.

Whilst proactive design choices may counter the severing of the human link, some choices inflame another debate that rages intermittently within graphic design – that of style versus substance – essentially a debate about surface and depth. For some, style is the equivalent of graphic design's four-letter word, seen as 'false, shallow and meaningless' (Blauvelt 1995: 64). However, this ignores a 'communicative code' that distinguishes between readers of different cultural and social groups (Blauvelt 1995). Style is therefore part of the rhetoric, and cannot be divorced from content – the dualism is an artificial construction – 'style is content too' (Bruinsma 1999: 2). Rock (2009) goes as far as to pronounce '*Fuck Content*', arguing that 'the materiality of a designer's method is his or her content' and how the designer 'speaks'. Style is a critical part of a designer's toolkit, but adopting a style simply because of fashion, and employing it inappropriately is how style has become so misunderstood.

Conversely, the modernist designer's aim was to relay a message clearly and concisely, without extraneous ornament, and to produce a universal visual language. This portrays the designer, and the typography, as purely objective (Armstrong 2009: 11; Blauvelt 1995: 65), though as Kinross (1985) points out this is actually 'a rhetoric of neutrality'. By the mid 1950s, modernist typography had been widely adopted by large corporations, particularly in the United States, and the 'utopian ideal of a universality of form and visual language' became a style of which the main features were white space, asymmetrical layouts and sans serif typography (Noble & Bestley 2001: 32). Today, the modernist 'style' is still regularly used – *Helvetica Neue* is considered fashionable and systems, information graphics and white space are prevalent. Keedy (2003: 59) described this twenty-first century version of modernism as 'Modernism 8.0', the 'latest upgrade' since Modernism's inception 80 years ago. He suggests it provided a refuge from the complications of postmodernism and an easy style guide for 'clueless' designers. This is a harsh critique, but Keedy wishes to further the discipline and a retreat to 'fundamentally conservative dogma' (1993: 29) denies typography much of its ability to question and provoke critical debate. However, the debates on style suggest there is more to *Zombie Modernism* (Keedy 1995) than meets the eye. The adoption of the modernist typographic style reveals it as a craft in itself, one that doesn't purvey a sinister rhetoric of neutrality, but one that speaks directly to its new target audience – the style conscious twenty-something.

Graphic design and typography: from theory to practice

Typography has engendered vivid verbal and visual exchanges surrounding notions of style, ideology, and theoretical exploration. The traditional practice of typography is underpinned by a series of rules, or conventions, that centre on the production of clear, legible texts. For Morison, writing in 1930:

> Typography is the efficient means to an essentially utilitarian and only accidentally aesthetic end, for enjoyment of patterns is rarely the reader's chief aim. Therefore, any disposition of printing material which, whatever the intention, has the effect of coming between author and reader, is wrong.
>
> (Morison 1930: 170)

Such rules are an anathema to postmodern typographers who don't subscribe to 'universally applicable values or solutions' (Poynor 2003: 11), and the 'sacred cow' of legibility has been 'assaulted' by new typographers such as Jeffery Keedy, in the move to dispel the myth of typographic neutrality (Poynor 1994: 84). Much of the new typography emanated from Cranbrook Academy in the United States where students were engaging with poststructuralist theoretical texts, in particular those by Derrida, and making connections with their visual work. This led them to attempt to: 'Deconstruct or break apart and expose, the manipulative visual language and different levels of meaning embodied in design in the same way that a literary critic might deconstruct and decode the verbal language of a novel' (Poynor 1991: 13). This was an attempt to probe beneath the surface of the page and engage with language and meaning through the medium of typography. As such, it was the first time graphic designers had engaged with such philosophical and theoretical positions from outside the discipline and Poynor suggests that graphic design has 'long had an aversion to theory' (2003: 10). Graphic design was only coined as a term in 1922 (Livingston & Livingston 1992: 59) and the profession was not fully established until the middle of the twentieth century (Hollis 1994: 8). In the UK, the first degree courses in polytechnics were not offered until the mid 1960s and it wasn't until 1992, when polytechnics received university status, that graphic design was taught as a university subject. In academic terms, therefore, graphic design does not have the well-established research territory of subjects such as geography. This aversion to theory could perhaps be defended because of the relative immaturity of the discipline, but many retrospective critiques of the 'new typography' found it to be flawed. Some designers did attempt to engage with the ideas on a deeper level (Abbott Miller & Lupton 1996; Byrne & Witte 1990; Lupton 1994), and during the early 1990s design and typography magazines such as *Emigre* and *Eye* provided a platform for such work which developed into a largely unresolved debate about graphic authorship (Poynor 2003: 144). Rock's (1996) article, *The designer as author*, attempted to address the contradictory issues of the nature of authorship in relation to the context of graphic design and he remains sceptical of the use of the term. However, the article was mistaken by many graphic designers for an invitation to produce work that was criticised by many as at best self-initiated, at worst, self-indulgent. Some critics now feel that although the design work expressed the poststructuralist theoretical ideas visually, many designers (much like Harley in the context of cartography) did not fully understand the texts they claimed to be inspired by (Drucker & McVarish 2009). For example, Barthes positions the role of the author as subservient to that of the reader.

> . . .a text is made of multiple writings drawn from many cultures and entering into mutual relations of dialogue, parody, contestation, but there is one place where this multiplicity is focused and that place is the reader. The reader is the space on which all the quotations that make up a writing are inscribed without any of them being lost; a text's unity lies not in its origin but in its destination.
>
> (Barthes 1977: 148)

Barthes (1977: 142) further states that 'as soon as a fact is narrated . . . the voice loses its origin, the author enters into his own death, writing begins'. So the debate about graphic authorship actually exists in contradiction to the theories allegedly underpinning it. Much of this early experimental typographic 'authorial' work now rests in monographs of designers elevated to hero status – for example, David Carson (Blackwell 1995) and Neville Brody (Wozencraft 1994) – and the work has ultimately become known and revered for its visual style, rather than content or message.

However, at the time, neither aesthetic nor theoretical approval of such experimental work was universally forthcoming. Rand (1992) accuses designers of being obsessed with theory and using it to generate impenetrable visual language and new style for the sake of it. This response, from an arch-modernist, drew many supporters. In *Cult of the Ugly* (1993), Heller derided such work as 'driven by instinct and obscured by theory, with ugliness its foremost by-product' (1993: 155), and suggested that it would be a 'blip . . . in the continuum of graphic design history' (1993: 158). Heller also feared that these new typographic experiments in design schools would be consumed by commercial designers and used as 'style without substance' (Poynor 2003). Ironically, Heller later simplistically reduces such work to style himself, and states that 'now postmodernism, deconstruction and grunge are history, the next new thing has yet to emerge' (Heller 2002: x), and both Blauvelt (2003) and Novosedelik (1996) note that self-expression rapidly replaced any real critical exploration. Fitzgerald (2003: 16) states that it is no surprise that these 'non-conformist forms' were absorbed into the mainstream as this is simply the 'life cycle of style'. So what began as a theoretical exploration of typography, language and graphic design in academia, became a commercially used style to target the youth market. Ultimately, and somewhat ironically, a backlash to 'decon' led to the above reincarnation of the modernist style. This manoeuvring from one 'style' to another stems from simplistic oppositional thinking about terms such as modernism and postmodernism.

A more ambitious agenda

So, could it be possible to sidestep this repetitive 'life cycle of style' and engage beyond the 'surface' of the page? The two camps within graphic design have been described by Bell (2004) as 'agents of neutrality' and 'aesthetes of style', but Beirut (2004) calls for a third way, one that suggests design should be a response to content. Mermoz (1995; 1998; 2000; 2002; Mau & Mermoz 2004) has provided a more theoretically led challenge to designers to fully engage with the possibilities of a content driven approach to typography. Critical of what he described as the 'retinal' state of graphic design and typography (1998, 2002, 2006), Mermoz also seeks to redefine the debate beyond 'surface pattern and complacent self-expression' (1998: 141). To do this, Mermoz proposes to

> conceptualise typography as a structure generative of meanings which, working in conjunction with other systems—such as the code of language, rhetoric and stylistics—is capable not just of making a text legible, readable and/or visually attractive; but also, by working at the level of the text, of reinforcing the strategy of the text and extending its effects beyond mere visual appearance and ergonomic function.
>
> (Mermoz 2002: 287)

In this sense, the text is defined as a 'semiotic object, rather than a mere physical entity' (Mermoz 2002: 287; see also van Leeuwen 2006: 144), with the typographic text working at two levels of denotation – the first being legibility, the second 'meanings and interpretation/s' (Mermoz 1995: np). This, suggests Mermoz, is 'typography as language in performance', but this is a performance that is at the service of the text, not one that engages in visual typographic pyrotechnics that obscure meaning – 'it is not from gazing at the page, but from reading the text that typography reveals its functionalities' (Mermoz 1998: 43). The role of the typographer 'is to make explicit the strategy of the text; the processes through which it is implemented . . . and . . . the plurality of meanings elicited from them' (Mermoz 2002: 290). This proliferation of meanings brings the work of the typographer to a point where it converges with Barthes' (1990: 4) definition of reading and the notion of the writerly text. The goal of the writerly text is to position 'the reader no longer as a consumer, but as a producer of the text', with the writerly text plural in its interpretations and meanings. The idea of the open, writerly text, along with the lexia – a fragment of text that carries as many as three or four meanings (Barthes 1990: 13) – has also been used widely within hypertext; screen-based writing that functions in an interactive environment offering the reader a 'multi-linear or multi-sequential experience'. Hypertext 'blurs the boundaries between reader and writer' (Landow 1997: 4), enabling a new reading on each engagement, aligning with Massey's place as process and echoing Kitchin & Dodge's (2007) conception of mapping as process and the 'map spaces' of Del Casino & Hanna (2006). The prospect of a text reclaiming the sense of place that Ingold (2007) alleges is lost, and the reader being able to explore it in a non-linear fashion, is a distinct possibility with this open, writerly approach.

Summary

The rise in the use of creative research methods and art-geography collaborations has been driven by non-representational thinking and a return to phenomenological approaches that seek an embodied, experiential understanding of place. However, at present there seems to be something of a gap between these experimental methods and the traditional journal articles that are produced to disseminate them. There also seems to be a focus on methods that might be said to develop re/presentations of place that are more fluid and less fixed than the printed word – for example, film, sound, or the spoken word. To this end, the creative turn within geography and many other disciplines has largely been framed within the context of art, with design rarely mentioned as having the potential to contribute to methodological developments. Yet, graphic design and typography have the potential to develop responses to place that are simultaneously both text based *and* visual. The use of creative research methods is still an emergent field, and questions remain about specific approaches that require more explicit discussion by those who undertake them (Pink 2015: xii-xiii: Davies & Dwyer 2010: 95). Questions also remain as to how to evaluate such work. As we have seen in Springer's (2017) call for geopoetics and Butz's (2011) concerns over publishing the *Bus Hub*, the majority of work is assessed using strict criteria developed for a traditional academic journal article. So it is perhaps not surprising the results produced via the use of such creative methods are open to question as to whether they 'should be held to account aesthetically and/or intellectually and should be deemed both art and geography' (Hawkins 2012: 65).

Perhaps this is an indication that in some cases there is a sense that the work is one but not the other, with various geographers questioning whether disciplinary boundaries can be blurred successfully (Nash 2013: 52) and whether artists involved are fully immersed in a geographic understanding of everyday life and place (Tolia-Kelly 2012: 137; Jellis 2015: 369). This book is an attempt to cross many of these divides, and the following chapter outlines a methodological approach that draws from both ethnography and design. In doing so it enables an engagement with everyday life and place that encourages interdisciplinarity and the production of research that can be evaluated in terms of its intellectual *and* aesthetic contributions. In doing so it sees these less as separate entities rather as inextricably and productively linked in the process and production of the work.

Bibliography

Abbott Miller, J. & Lupton, E. (1996) 'Deconstruction and graphic design' in Abbott Miller, J. & Lupton, E. *Design Writing Research*. New York: Princeton Architectural Press, pp. 3–23.

Armstrong,H. (2009) 'Introduction: Revisiting the avant-garde' in Armstrong, H. (ed.) *Graphic Design Theory: Readings From the Field*. New York: Princeton Architectural Press, pp. 9–15.

Atkinson, M. (1990) *The Ethnographic Imagination: Textual Constructions of Reality*. London: Routledge.

Attoh, K. (2011) 'The bus hub', *ACME: An International Journal for Critical Geographies*. 10(2), pp. 280–285.

Barnes, A. (2012) 'Thinking geo/graphically: The interdisciplinary space between graphic design and cultural geography', *Polymath: An Interdisciplinary Arts and Sciences Journal*. 2(3), pp. 69–84.

Barthes, R. (1990) *S/Z*. London: Blackwell Publishing.

Barthes, R. (1977) *Image, Music, Text*. London: Fontana.

Beirut, M. (2004) 'The world in two footnotes', *Design Observer*. Available at: http://design observer.com/feature/the-world-in-two-footnotes/2637 (Accessed: 30 December 2017).

Bell, N. (2004) 'The steamroller of branding', *Eye*. 53(14) Available at: www.eye magazine.com/feature/article/the-steamroller-of-branding-text-in-full (Accessed: 30 December 2017).

Benjamin, W. (1999) *The Arcades Project*. Cambridge, Mass.: Belknapp Press.

Blackwell, L. (1995) *The End of Print: The Graphic Design of David Carson*. London: Laurence King.

Blauvelt, A. (2003) 'Untitled', *Emigre*. 64, pp. 36–43.

Blauvelt, A. (1995) 'Under the surface of style', *Eye*. 18, pp. 64–70.

Brace, C. & Johns-Putra, A. (2010) 'Recovering inspiration in the spaces of creative writing', *Transactions of the Institute of British Geographers*. 35(3), pp. 399–413.

Bruinsma, M. (1999) 'Style is content', *Eye*. 32, p. 2.

Buchanan, R. (1989) 'Declaration by design: rhetoric, argument and demonstration in design practice', in Margolin, V. (ed.) *Design Discourse*. Chicago: Chicago University Press, pp. 91–109.

Butz, D. (2011) 'The bus hub: Editor's preface', *ACME: An International Journal for Critical Geographies*. 10(2), pp. 278–279.

Byrne, C. & Witte, M. (1990) 'A brave new world: Understanding deconstruction' in Beirut, M., Drenttel, W., Heller, S., & Holland, D. K. (eds) *Looking Closer: Critical Writings on Graphic Design.* New York: Allworth Press, pp. 115–121.

Cameron, E. (2012) 'New geographies of story and storytelling', *Progress in Human Geography.* 36(5), pp. 573–592.

Coles, B. F. (2014) 'Making the market place: a topography of Borough Market, London', *cultural geographies.* 21(3), pp. 515–523.

Cresswell, T. (2015) *Place: An Introduction.* 2nd edn Chichester: John Wiley & Sons.

Cresswell, T. (2013) 'Displacements: Three poems', *The Geographical Review.* 103(2), pp. 285–287.

Daniels, S. & Lorimer, H. (2012) 'Until the end of days: narrating landscape and environment', *cultural geographies.* 19(1), pp. 3–9.

Daniels, S., Pearson, M. & Roms, H. (2010) 'Editorial', *Performance Research.* 15(4), pp. 1–5.

Davies, G. & Dwyer, C. (2010) 'Qualitative methods III: animating archives, artful interventions and online environments', *Progress in Human Geography.* 34(1), pp. 88–97.

Davies, G. & Dwyer, C. (2007) 'Qualitative methods: Are you enchanted or are you alienated?', *Progress in Human Geography.* 31(2), pp. 257–266.

de Leeuw, S. (2015) *Skeena.* British Columbia: Caitlin Press.

de Leeuw, S. (2012) *Geographies of a Lover.* Edmonton: NeWest Press.

de Leeuw, S. (2011) 'New Routes of Geographic Contemplation: Poetry and Public Transportation', *ACME: An International Journal for Critical Geographies.* 10(2), pp. 289–292.

Del Casino, V. J. & Hanna, S. P. (2006) 'Beyond the "binaries": A methodological intervention for interrogating maps as representational practices', *ACME: An International Journal for Critical Geographies.* 4(1), pp. 34–56.

DeSilvey, C. (2012) 'Making sense of transience: an anticipatory history', *cultural geographies.* 19(1), pp. 31–54.

DeSilvey, C. & Yusoff, K. (2006) 'Art and geography' in Douglas, I., Hugget, R., & Perkins, C. (eds) *Companion Encyclopaedia of Geography.* London: Routledge, pp. 573–588.

Driver, F., Nash, C. & Prendergast, K. (2002) *Landing: Eight Collaborative Projects Between Artists and Geographers,* Egham: Royal Holloway, University of London.

Drucker, J. & McVarish, E. (2009) *Graphic Design History: A Critical Guide.* New Jersey: Prentice Hall.

Eco, U. (1989) *The Open Work.* Cambridge, Mass.: Harvard University Press.

Eshun, G. & Madge, C. (2016) 'Poetic world-writing in a pluriversal world: a provocation to the creative (re)turn in geography', *Social and Cultural Geography.* 17(6), pp. 778–785.

Fitzgerald, K. (2003) 'Quietude', *Emigre.* 64, pp. 16–32.

Foster, K. & Lorimer, H. (2007) 'Cultural geographies in practice: Some reflections on art-geography as collaboration', *cultural geographies.* 14(3), pp. 425–432.

Frascara, J. (2006) *Designing Effective Communications: Creating Contexts for Clarity and Meaning.* New York: Allworth Press.

Gallagher, M. and Prior, J. (2014) 'Sonic geographies: Exploring phonographic methods', *Progress in Human Geography.* 38(2), pp. 267–284.

Gibson-Graham, J. K. (2014) 'Rethinking the economy with thick description and weak theory', *Current Anthropology.* 59(S9), pp. S147–S153.

Gibson-Graham, J. K. (2008) 'Diverse economies: performative practices for "other worlds"', *Progress in Human Geography.* 32(5) pp. 613–632.

Hawkins, H. (2017) *Creativity*. Abingdon: Routledge.

Hawkins, H. (2012) 'Geography and art. An expanding field: Site, the body and practice', *Progress in Human Geography*. 37(1), pp. 52–71.

Heller, S. (2002) 'Introduction' in Beirut, M., Drenttel, W. & Heller, S. (eds) *Looking Closer 4: Critical Writings on Graphic Design*. New York: Allworth Press, pp. ix–xi.

Heller, S. (1993) 'Cult of the Ugly', in Beirut, M., Drenttel, W., Heller, S., & Holland, D. K. (eds) (1994) *Looking Closer: Critical Writings on Graphic Design*. New York: Allworth Press, pp. 155–159.

Hollis, R. (1994) *Graphic Design: A Concise History*. London: Thames & Hudson.

Ingold, T. (2007) *Lines: A Brief History*. Abingdon: Routledge.

Jellis, T. (2015) 'Spatial experiments: art, geography, pedagogy', *cultural geographies*. 22(2), pp. 369–374.

Keedy, J. (2003) 'Modernism 8.0', *Emigre*. 64, pp. 58–71.

Keedy, J. (1995) 'Zombie Modernism' in Heller, S. & Meggs, P. (eds) (2001) *Texts on Type: Critical Writings on Typography*. New York: Allworth Press, pp. 159–167.

Keedy, J. (1993) 'The rules of typography according to crackpots experts' in Beirut, M., Drenttel, W., Heller, S., & Holland, D. K. (eds) (1997) *Looking Closer 2: Critical Writings on Graphic Design*. New York: Allworth Press, pp. 27–31.

Kinman, E. & Williams, J. (2007) 'Domain: Collaborating with clay and cartography', *cultural geographies*. 14(3), pp. 433–444.

Kinross, R. (1985) 'The rhetoric of neutrality, *Design Issues*. 2(2), pp. 18–30.

Kitchin, R. & Dodge, M. (2007) 'Rethinking Maps', *Progress in Human Geography*. 31(3), pp. 331–344.

Landow, G. (1997) *Hypertext 2.0*. Baltimore: The John Hopkins University Press.

Livingston, A. & Livingston, I. (1992) *The Thames & Hudson Dictionary of Graphic Design and Designers*. London: Thames & Hudson.

Lorimer, H. (2008) 'Poetry and place: The shape of words', *Geography*. 93(3), pp.181–182.

Lorimer, H. (2005) 'Cultural geography: The busyness of being "more than representational"', *Progress in Human Geography*. 29(1), pp. 83–94.

Lorimer, H. (2003) 'Telling small stories: spaces of knowledge and the practice of geography', *Transactions of the Institute of British Geographers*. 28(2), pp. 197–217.

Lorimer, H. & MacDonald, F. (2002) 'A rescue archaeology, Taransay, Scotland', *cultural geographies*. 9(1), pp. 95–102.

Lorimer, J. (2010) 'Moving image methodologies for more-than-human geographies, *cultural geographies*. 17(2), pp. 237–258.

Lovejoy, A. & Hawkins, H. (2009) *Insites: An Artists' Book*. Penryn: Insites Press.

Lupton, E. (1994) 'A post-mortem on deconstruction' in Heller, S. & Meggs, P. (eds) (2001) *Texts on Type: Critical Writings on Typography*. New York: Allworth Press, pp. 45–47.

Marston, S. & De Leeuw, S. (2013) 'Creativity and geography: Toward a politicized intervention', *The Geographical Review*. 103(2), pp. iii–xxvi.

Massey, D. (2005) *For Space*. London: Sage.

Massey, D. (1994) *Space, Place and Gender*. Minneapolis: University of Minnesota Press.

Mau, B. & Mermoz, G. (2004) 'Beyond looking: Towards reading. . .', *Baseline*. 43, pp. 33–36.

Mermoz, G. (2006) 'The designer as author: Reading the city of signs—Istanbul: Revealed or mystified?, *Design Issues*. 22(2), pp. 77–87.

Mermoz, G. (2002) 'On Typographic Signification. . .', *Mind the Map: Third International Conference on Design History & Design Studies*, Istanbul, Turkey, 9–12 July, pp. 287–290.

Mermoz, G. (2000) 'Towards more ambitious agendas' in Mealing, S. & Dudley, E. (eds) *Becoming Designers*. Exeter: Intellect, pp. 152–158.

Mermoz, G. (1998) 'Deconstruction and the typography of books', *Baseline*. 25, pp. 41–44.

Mermoz, G. (1995) 'On typographic reference: Part 1, *Emigre*. 36, no pagination.

Morison, S. (1930) 'First principles of typography' in Heller, S. & Meggs, P. (eds) (2001) *Texts on Type: Critical Writings on Typography*. New York: Allworth Press, pp. 170–177.

Nash, C. (2013) 'Cultural geography in practice' in Johnson, N., Schein, R. H. & Winders, J. (eds) *The Wiley-Blackwell Companion to Cultural Geography*. Chichester: John Wiley & Sons, pp. 45–56.

Nash, C. (2000) 'Performativity in practice: Some recent work in cultural geography', *Progress in Human Geography*. 24(4), pp. 653–664.

Noble, I. & Bestley, R. (2001) *Experimental Layout*. Hove: RotoVision.

Novosedelik, W. (1996) 'Dumb', *Eye*. 15(4), pp. 54–57.

O'Neill, M. (2008) 'Transnational refugees: The transformative role of art?, *Forum: Qualitative Social Research*. 9(2), Art. 59.

O'Neill, M. & Hubbard, P. (2010) 'Walking, sensing, belonging: ethno-mimesis as performative practice', *Visual Studies*. 25(1), pp. 46–58.

Parr, H. (2007) 'Collaborative filmmaking as process, method and text in mental health research', *cultural geographies*. 14(1), pp. 114–38.

Pink, S. (2015) *Doing Sensory Ethnography*. London: Sage.

Poynor, R. (2003) *No More Rules: Graphic Design and Postmodernism*. London: Laurence King.

Poynor, R. (1994) 'Type and deconstruction in the digital era' (revised edition) in Beirut, M., Drenttel, W., Heller, S., & Holland, D. K. (eds) *Looking Closer: Critical Writings on Graphic Design*. New York: Allworth Press, pp. 83–87.

Poynor, R. (1991) 'Type and deconstruction in the digital era' in Poynor, R. & Booth-Clibborn, E. (eds) *Typography Now: The Next Wave*. London: Booth-Clibborn Editions, pp. 6–19.

Price, P. L. (2004) *Dry Place: Landscapes of Belonging and Exclusion*. Minneapolis: University of Minnesota Press.

Pryke, M. (2002) 'The white noise of capitalism: Audio and visual montage and sensing economic change', *cultural geographies*. 9(4), pp. 472–477.

Rand, P. (1992) 'Confusion and chaos: The seduction of contemporary graphic design' in Heller, S. & Finamore, M. (eds) (1997) *Design Culture: An Anthology of Writing From the AIGA Journal of Graphic Design*. New York: Allworth Press, pp. 119–124.

Roberts, B. (2008) 'Performative social science: A consideration of skills, purpose and context', *Forum: Qualitative Social Research*. 9(2), Art. 58.

Rock, M. (2009) *Fuck Content*. Available at: https://2x4.org/ideas/2/fuck-content/ (Accessed: 30 December 2017).

Rock, M. (1996) 'The designer as author' in Beirut, M., Drenttel, W. & Heller, S. (eds) (2002) *Looking Closer 4: Critical Writings on Graphic Design*. New York: Allworth Press, pp. 237–244.

Rosnau, L. (2013) 'Four poems', *The Geographical Review*. 103(2), pp. 139–142.

Ryan, J. (2003) 'Who's afraid of visual culture?', *Antipode*. 35(2), pp. 232–237.

Schaaf, R., Worrall-Hood, J. & Jones, O. (2017) 'Geography and art: encountering place across disciplines', *cultural geographies*. 24(2), pp. 319–327.

Springer, S, (2017), 'Earth Writing', *GeoHumanities*. 3(1), pp. 1–19.

Thrift, N. (2000) 'Dead or alive?', in Cook, I., Crouch, D., Naylor, S., & Ryan, J. (eds) *Cultural Turns/Geographical Turns: Perspectives on Cultural Geography*. Harlow: Prentice-Hall, pp. 1–6.

Tolia-Kelly, D. (2012) 'The geographies of cultural geography II: Visual culture', *Progress in Human Geography*. 36(1), pp. 135–142.

Tyler, A. (1992) 'Shaping belief: The role of the audience in visual communication', *Design Issues*. 9(1), pp. 21–29.

van Leeuwen, T. (2006) 'Towards a semiotics of typography', *Information Design Journal*. 14(2), pp. 139–155.

Wigmore, G. (2013) 'Black rocks and sea', *The Geographical Review*. 103(2), pp. 256–259.

Wozencraft, J. (1994) *The Graphic Language of Neville Brody 2*. London: Thames & Hudson.

Wylie, J. (2006) 'Smoothlands: Fragments/landscapes/fragments', *cultural geographies*. 13(3), pp. 458–465.

4 Understanding everyday life and place

Introduction

The previous three chapters have outlined the theoretical perspectives a geo/graphic design process draws on and in this chapter we move to discuss the methodological concerns that together provide the framework for work of a geo/graphic nature. Interdisciplinary research naturally takes a multi-method approach, one that has been described as that of the bricoleur (Denzin & Lincoln 2005: 4), the researcher who uses a set of interpretive practices 'tailored to the individual project' and 'responsive, driven by the requirements of practice' (Gray & Malins 2004: 72). However, these methods are not chosen haphazardly, but are taken from interlinked and related approaches in order to form a 'developmental set, which is coherent' (Gray & Malins 2004: 72–74). This approach facilitates the construction of a bricolage, 'a complex . . . reflexive, collage or montage—a set of fluid, interconnected images and representations' (Denzin & Lincoln 2005: 6). A bricolage 'recognises the dialectical nature' of such an interdisciplinary study, and 'promotes a synergistic interaction' (Kincheloe 2001: 679). Here, the construction of a bricolage through the deployment of various methods from the social sciences and graphic design has enabled the development of a 'new methodological synthesis' (Kincheloe 2001: 685). In research such as this it is in the 'liminal zones where disciplines collide' that new approaches are likely to be found (Kincheloe 2001: 689; see also Rendell 2006: 11).

Such an interdisciplinary space has also been described by Pearson & Shanks (2001) as creating a 'blurred genre', where two disciplines are no longer functioning as discrete entities. Such a blurred genre necessitates 'different ways of telling and different types of recording and inscription, which can incorporate different orders of narrative. It suggests mutual experiments with modes of documentation which can integrate text and image . . .' (Pearson & Shanks 2001: 131). The term they use for such ways of telling is 'incorporations'. An incorporation is a site report that acknowledges

> Juxtapositions and interpenetrations of the historical and contemporary, the political and the poetic, the factual and fictional, discursive and the sensual. These are proactive documents: their parts do not necessarily cohere. They will require work but they leave space for the imagination of the reader.
>
> (Pearson & Shanks 2001: 159)

The specific methodological approaches and methods drawn on to develop the geo/graphic design process include ethnography and, more specifically, visual and sensory ethnography alongside the practice of walking; writing as a method of inquiry; design as a method of inquiry; and 'noticing, collecting and thinking' (Seidel 1998), prototyping and reflection as methods of analysis.

Ethnographic methods of experiencing and understanding place

If one wishes to understand and represent place one needs to experience place in person rather than solely rely on secondary sources from the comfort of one's desk. The use of ethnographic methods enables a direct connection between the researcher, place and its representation – an immediate encounter with 'real world messiness' (Crang & Cook 2007: 14). Ethnography is therefore not just about what people say concerning their everyday life, it's also about what they actually do.

Participant observation, reflexivity and the position of the researcher

In order to study the specific research site and participants in depth, ethnographers traditionally spend an extended period of time 'in the field'. In this way, an ethnographer becomes embedded within the particular cultural group that is being studied and is therefore able to develop an understanding of everyday life in that context. Many research methods handbooks distill ethnography into a set of methods, most prominently participant observation and interviewing. Hammersley and Atkinson describe this set of methods as involving

> the researcher participating, overtly or covertly, in people's daily lives for an extended period of time, watching what happens, listening to what is said, asking questions—in fact, collecting whatever data are available to throw light on the issues that are the emerging focus of inquiry.
>
> (2007: 1)

Participant observation is focused on gaining an understanding of the space of the research via the retention of an outsider's perspective, with the researcher needing to maintain some kind of distance. Clifford (1998: 34) describes this process as a 'continuous tacking between the "inside" and "outside" of events'. Retaining this perspective is vital, and any sense that the researcher is beginning to feel at home or get overly comfortable in their surroundings suggests that perspective may be beginning to be lost. The researcher should always strive to maintain both a social and an intellectual distance in order to enable the process of analysis (Hammersley & Atkinson 2007: 90). Such references to inside and outside and maintaining distance, may inevitably conjure up notions of 'the Other' and dated colonial practices of exploration. However, as we will discuss below, contemporary ethnography and ethnographers are reflexive, aware of the potential power dynamic inherent in the relationship between researcher and the researched, and engage people as participants rather than position them as subjects. So, rather than using the idea of distance to position the ethnographer as somehow superior to those who are part of the study, it is simply a strategy for trying to defamiliarise the type of everyday activities that we might otherwise take for granted. Highmore (2002: 87) goes as far as to suggest that practicing ethnography 'at home' enables its most 'critical possibilities' to be revealed. So, if we are observing something that is familiar to us we need to 'step back' from it in order to reflect on it as if it were new to us. For example, waiting for a bus is something that happens regularly across the world, however, in the UK people not only wait, but they queue. Someone used to queuing for their bus to work in a morning, would give this little thought, but if we interrogate the practice of queuing in this context a little more we will begin to understand more about everyday life and place. For example, how does the queue form, what happens if someone disregards the unwritten rules, how do people use their bodies to maintain their position in the queue, and what kind of words or gestures, if any, are exchanged if the system breaks down? Queuing is something of a ritual in the UK and ritual elements that constitute everyday life and place can be taken as culturally significant, but it is only through this deeper questioning that the significance is revealed as it is rarely self-evident or easily interpreted (Highmore 2002: 88).

In the context of participant observation, ethnographers keep field notes or diaries. These contain everything from notes taken quickly whilst observing something unfold, to longer, more reflective pieces written after the event. A range of texts offer suggestions for how to approach field notes and combine various layers of description in order build a detailed picture of the site of study (for example Crang & Cook 2007: 51; Cloke et al. 2004: 200). Whilst such methods offer a very clear framework to take into the field, often, contemporary settings are not 'bounded' in such a way as to make this approach feasible. It also shouldn't be assumed that 'doing ethnography' is simply a case of describing and analysing a setting in writing. Whilst ethnographic practice often centres on participant observation and interviewing, ethnographers use a wide range of collaborative strategies. They also often adapt and develop these in the course of the research depending on the different contexts and participants and because of this Pink suggests there is 'now no standard way of doing ethnography that is universally practiced' (Pink 2015: 4). One alternative recording strategy is that of 'the salience hierarchy' (Wolfinger 2002: 89). Here the ethnographer simply starts by describing what strikes them as the most interesting observation within the setting. A similar strategy is that of 'floating observation' which 'consists in keeping one's responsiveness, not focusing one's attention upon any specific object' and allowing chance encounters to emerge (Petonnet 1982: 47). Whichever strategy one uses, the notes will inevitably be informed by the researcher's tacit knowledge, beliefs and experience (Wolfinger 2002: 93). Therefore one must subject immediate assumptions to a process of reflection and further observation.

Thus, in reflecting on any event and framing it within our own understandings and experiences, the positioning of the researcher is central to this knowing. As we have already seen in Chapter 2, the 'crisis of representation' challenged the belief that an objective, authentic sense of everyday life and place can be communicated. This led to the introduction of the idea that any ethnographic research inevitably produces intersubjective, 'inherently partial' truths (Clifford 1986: 7, Duncan & Ley 1993: 4). Essentially we all see the world through our own eyes, and therefore construct our own 'truths'. Our perspective will be affected by a range of factors, for example, gender, sexuality, race, ethnicity, or class, and this is the same for both the researcher and the participants. Therefore, ethnographers began to look more critically at how ethnographic research was being undertaken and written up. In doing so, they revealed that a clear power difference was evident in the majority of ethnographic work conducted to this point, with those undertaking the study either members of a colonial power or higher class, and those being studied either colonised or from a lower class. In this context, the subjects of the study were also more often than not portrayed as either exotic, or lacking in sophistication or intelligence. Through analysing how ethnographies like this were constructed, those at the forefront of the reflexive turn realised that how language was used was integral to how the group under study was understood. Therefore, the realisation that rhetoric was central to perception led to ethnographies being seen as 'fiction' (Clifford 1986: 6).

The response to this realisation was to develop a postmodern perspective on ethnography that attempts to remove power from the individual ethnographer and develop more inclusive accounts of everyday life and place. These reflect the complex, multi-layered, relational perspectives inherent in contemporary versions of place and challenge the previously held view that there is a metanarrative waiting to be discovered. In order to achieve this, many ethnographers began to develop writing and other representational strategies that enabled multiple and different perspectives to be heard. For some, postmodern approaches are too relativistic, in danger of abandoning a sense of any kind of 'real world'. This has ultimately resulted in a 'reflexive turn', one that accepts that as researchers we are embedded in the world that we are studying and that what we choose to observe, and our analysis of these observations, will inevitably be reflected and refracted through our own life experiences. A reflexive approach therefore attempts to take into account an awareness of this personal position and an understanding of the potential power differences between the ethnographer and participants. Those under study are positioned as participants and engaged in conversation, often also participating in decisions as to how the research is undertaken and presented − this endeavours to close the gap between the researcher and the researched. Reflexive ethnographers also contextualise their research and their place in it, in an attempt to give the reader a sense of their perspective and who or what may have influenced the trajectory of the project. However, reflexivity should not be seen as a way of removing bias, rather, subjectivity is 'engaged with as a central aspect of ethnographic knowledge, interpretation and representation' (Pink 2013: 36). The position of the resultant ethnography is, therefore, a partial view, one story amongst many that make up the complexity of place, but it is one in which the intersubjective production of that story between researcher, participants and place is made apparent (Pink 2013: 36). However, Pink is at pains to stress that simply because it doesn't claim to be objective, any reflexive ethnography should still offer an account of the ethnographer's 'experiences of reality that are as loyal as possible to the context, the embodied, sensory and affective experiences' as well as 'the negotiations through which the knowledge was produced' (Pink 2013: 35).

Whilst many of the more experimental contemporary methods used in geography today have a direct legacy to ethnography (Shaw, DeLyser & Crang 2015: 212), ethnography is more than a set of methods, and can be defined as a methodology. This is something Pink expands on in the context of her work on both visual and sensory ethnographies.

Visual and sensory ethnography

The idea of a visual ethnography perhaps seems obvious – we are constantly told that 'images are everywhere' and a camera has always been part of an ethnographer's tool kit. Until the late 1980s, the ethnographic literature relating to photography was that of the 'how-to' manual and positioned it as unmediated and capable of offering an objective, realist image of everyday life and place. However, as we have seen above, with the advent of postmodern thinking such approaches to knowledge were challenged and this led to the recognition that an image was no more or less subjective than a written text (Pink 2013: 3). Therefore, ethnographers began to explore the creation and use of imagery as an ethnographic method in its own right. In this context, Pink's articulation of a visual ethnography is an attempt to understand and communicate how images (both still and moving) are implicated in both 'the production and dissemination of the ways of knowing that are part of the ethnographic process' (Pink 2013: 2). However, as with her assertion that there is no clear roadmap for how to undertake fieldwork, Pink also suggests that there are no fixed criteria that can be used to assess whether a photograph, or any other text or piece of information is ethnographic or not. As with every text, in its broadest sense, every photograph can hold different significance for, and be interpreted differently by, different people at different times (Pink 2013: 35). Photographic images are therefore perhaps best seen as 'representations of aspects of culture' (Pink 2007: 75). However, photographs taken as part of fieldwork can also act as triggers for reflexive thinking (Pink 2013: 85), therefore photography is a productive analytical tool in different contexts and at different stages of the ethnographic process. In proposing a visual ethnography, Pink is not arguing that images can be used to replace words, or play the same role as words might. Rather, she suggests that there should be no hierarchy of media within ethnographic representations, but that different epistemologies and media should complement each other. In this way, 'different types of ethnographic knowledge' may be 'experienced and represented in a range of different textual, visual and other sensory ways' (Pink 2013: 10).

In relation to the development of a sensory ethnography, Pink notes that, much like the 'spatial turn' and the 'practice turn', there is an increasing interest in the senses – she describes it as 'an explosion' – amongst scholars and practitioners within the social sciences and the arts and humanities (Pink 2015: xi). Taking an interdisciplinary approach to place, the multi-sensory experiences, perception, memory, knowing and practice, Pink proposes and develops an approach to undertaking and representing research that centres on sensory ways of experiencing everyday life and place. As we have seen in both the discussion of everyday life and that relating to non-representational theory, an embodied knowing through practice and multi-sensory experiences are integral to place and increasingly important to those who seek to understand and represent everyday life and place. However, sensory ethnography as proposed by Pink is not simply a set of methods to be applied alongside, or as part of, traditional methods such as participant observation or interviewing. Rather, Pink frames it (and visual ethnography) as a 'critical methodology', one that positions ethnography generally as 'a reflexive and experiential process through which academic and applied understanding, knowing and knowledge are produced' (Pink 2015: 3–4). Neither visual nor sensory ethnography subscribes to one particular research method or one particular type of data and, much like non-representational theory, both are open to new ways of working with the multi-sensory experiences of everyday life and place in order to develop new ways of knowing. Indeed, Pink is clear that 'data' has little part to play in this approach to ethnography – that rather than being about collecting data, it is about producing knowledge (Pink 2013: 31).

Ethnography is primarily associated with the disciplines of anthropology and sociology, however, it is used across a variety of disciplines and therefore inevitably reshaped by a range of 'disciplinary theories and priorities that inform the work it is required to undertake' (Pink 2013: 18). In a similar cross-disciplinary context, in developing the theoretical underpinnings for both a visual and sensory ethnographic approach in particular, Pink draws from a range of academic work. However, it should be pointed out that Pink offers up her writings on neither visual nor sensory ethnography as a closed definition or some kind of 'how to' manual. Rather, much like this book and ethnography generally, she offers them as a point of departure, one that others can bring their own disciplinary theories and practices to.

Knowing in practice

The concept of 'knowing in practice' (Wenger 1998: 141) is proposed by Pink as a way of thinking about how we might begin to experience and understand everyday life and place in a similar way to research participants (2015: 39). For Wenger, a 'community of practice' is characterised by three dimensions: mutual engagement, joint enterprise and shared repertoire (Wenger 1998: 73). One can see how this could relate to the idea of participant observation and to more collaborative, participatory methods of engagement that attempt to remove the gap between researcher and participants. By engaging in this community of practice we move towards an 'experience of knowing' as, for Wenger, 'knowing is defined only in the context of specific practices, where it arises out of the combination of a regime of competence and an experience of meaning' (1998: 142). Participation in everyday life and place is therefore key, with Wenger framing 'the experience of knowing' as one of 'participation' (1998: 142). Therefore, knowing is not some kind of removed, remote, purely cerebral analysis, it is 'experiential' (Wenger 1998: 141); it is embedded within the multi-sensory complexity of everyday life and place, and much like place, is 'constantly changing' (Pink 2015: 40). Thus, if knowing is a 'situated practice' (Pink 2015: 40), to experience, understand and 'know' place like research participants do, ethnographers need become an active participant themselves within this community of practice. For Pink (2015: 40), we can extend the idea of knowing in practice to an embodied, multi-sensory way of knowing that is emplaced and inextricable from material and sensorial interactions with the environment.

Ethnographic place-making

The definition of place used within this book conceptualises it as in a constant state of becoming, always emerging and forever unfinished. As every ethnographic researcher and participants in the research will themselves be emplaced, this raises the question of how the resulting ethnography and ethnographic representation of other emplaced persons are framed and conceptualised (Pink 2015: 48). Pink's idea of 'ethnographic places' is her solution to this, however, in this context, an ethnographic place is not the same as the site of the research, rather, it is 'the place that . . . ethnographers make when communicating . . . research to others' (Pink 2015: 48). This ethnographic place-making functions on three levels. The first level is place as it is experienced in the field. However, in the gathering of research materials and in the production of an ethnography the researcher engages in ethnographic place-making on the second level. Here, the materials that are being analysed act as triggers for memories of the research and our embodied experiences of place. Therefore, we 're-encounter the sensorial and emotional reality of research situations' and effectively create place anew (Pink 2015: 143). Whether an ethnographic representation includes a range of media or not, each one is the result of a weaving together of 'theory, discourse, memory and imagination' and can never be understood without accounting for how the reader or audience constitute meanings through their participation (Pink 2015: 48). Thus, when a reader or viewer engages with this representation, place is effectively remade on a third level. In this remaking, the reader or viewer bring their own experiences and imagination to the 'text' and therefore effectively create their own understandings and re-create place (Pink 2015: 125). This is reminiscent of Barthes' ideas of the 'death of the author' and 'birth of the reader', and perhaps simply extends postmodern revelations about the role of rhetoric in ethnographic writing. However, here, Pink is suggesting that images and texts could be constructed in such a way as to intentionally provoke an imaginative experience. For example, in a discussion of the use of photographic imagery in an ethnographic account of a journey, Pink suggests the reader makes their own journey through the text that is both 'experiential and imaginative'. In this context the images are not static photographs waiting for interpretation, they are part of the process of the ethnographic research, of the researcher's own journey through the project. To this end, they invite the reader to take that same journey and imagine the experiences and feelings the researcher encountered, thus the ethnography becomes an 'empathetic and experiential text as well as being a piece of scholarly work' (Pink 2013: 86). According to Pink, ethnographers should aspire to develop texts and representations of everyday life and place that engage their readers, allow them to imagine themselves in a particular scenario and result in some kind of impact or empathy (2015: 59). Indeed, Bakhtin (1981) suggests that effective narratives do just that, enabling a dialogic process by conveying information in a way that resonates with, and is further framed and extended by, the reader's experience. However, as we have seen in previous discussions on non-representational theory, this type of multi-sensory, embodied experience and knowing is difficult to express or capture in a traditional text-based response, and for Pink, this is one of the challenges of sensory ethnography (2015: 40).

Walking as an exploratory tactic of the everyday

As we have noted previously, ideas surrounding non-representational theory have led to the adoption of research methods that focus on the body, and geography itself is described as having undergone a 'bodily turn' (Hawkins 2017: 34). Walking as a research method for everyday life and place has come to the fore during this time. It continues to grow in popularity amongst academics and artists and several cultural geographers have focused on walking artists' urban exploration of the city (see Battista et al. 2005; Phillips 2005; Pinder 2001, 2005). To this end, it has been suggested that walking is not only 'fundamental to the everyday practice of social life', but also to 'much anthropological fieldwork' (Lee & Ingold 2006: 67). Walking has become 'increasingly central as a means of both creating new and embodied ways of knowing and producing scholarly narrative' (Pink et al. 2010: 1). The use of walking as a contemporary research method has its roots in the practice and writings of the Surrealists, de Certeau, Benjamin and the Situationist International, and ideas of the flâneur, the dérive and psychogeography.

The flâneur

The method of the flâneur was essentially to stroll and to observe – he is the 'secret spectator of the spectacle of the spaces and places of the city' and 'flânerie can be understood as the observation of the fleeting and transitory' (Tester 2015: 7). According to the poet Baudelaire, 'the crowd is his domain, just as the air is the bird's, and water that of the fish. His passion and his profession is to merge with the crowd' (Baudelaire 1995: 9). Many definitions of the flâneur reference words like idler or loafer, and in Italian, 'far flanella means "to walk around without any aim"' and is mainly attributed to unemployed people (Nuvolati 2014: 21). Deliberately contradictory terms such as 'alert reverie' and 'drifting purposefully' are often used to describe the flâneur (Nuvolati 2014: 23). By idling in an unobtrusive way rather than overtly observing, the flâneur becomes invisible to those in the crowd and essentially becomes one of them – 'a man of the crowd' – conscious of his role (Nuvolati 2014: 21–22). For Baudelaire, the flâneur is compelled to perform this role, it is not idling in the sense that he has nowhere else to be, this is 'doing' rather than simply 'being' (Tester 2015: 5). The purposeful nature therefore reflects the fact that it is a deliberate attempt to think of spaces and spatiality differently (Pile 2002: 212).

Benjamin, writing several decades after Baudelaire, was not entirely positive about the flâneur, he speculated that the rise of capitalism and increased commodification within the city left no spaces of mystery for the flâneur to observe (Nuvolati 2014: 22). He felt that the flâneur was a 'passive spectator' who was 'as duped by the spectacle of the public as the consumer who is duped by the glittering promises of consumerism' (Tester 2015: 14). It seems that Benjamin is reflecting on issues that also affect contemporary ethnographers and others who research everyday life and place. To conduct emplaced, embodied research inevitably means that one is both 'part of, and an active agent in the process' one is studying (Pink 2012: 31). We are all immersed in 'the everyday' every moment of our lives, so it perhaps comes with an added complication of being somehow less 'different' than a site one might have to travel long distances to, or a cultural context that is immediately unfamiliar. However, as we have discussed previously, undertaking research in a familiar context can be very productive (Highmore 2002: 87), but in order for this to occur researchers need retain something of an outsider's perspective along with adopting a reflexive approach that 'attends to the process of knowledge production, its intersubjectivity and the power relations that are embedded in it' (Pink 2012: 31). Although flânerie can be criticised for its privileged gaze and predominantly male focus, there is no doubt that the roots of much of this contemporary work reside in this nineteenth century Parisian practice.

Psychogeography and the dérive

By the late 1950s, interest in the everyday life, space, place and culture of the 'masses' had become popular within artistic circles in Europe and America (Sadler 1998: 11). During this time, the Situationist International, attempted to turn this avant-garde interest into a revolution. Profoundly influenced by the work of Henri Lefebvre, the Situationists were a political and artistic movement predominantly active in Paris between 1957 and 1972. They proposed ideas about urban living that were diametrically opposed to those of le Corbusier's vision of rational architecture that was prevalent in Paris at the time (Home 1996: 10). They called for the infilling of urban space, not its opening up, as they believed the production of excess space encouraged the capitalist circulation of things, which then trapped the inhabitants within it (Sadler 1998: 54). They identified the street as the space of real life in the city, and were determined 'to penetrate the outward, spectacular, commercialised signs of mass culture and explore its interior' (Sadler 1998: 19). The everyday patterns of life and place intrigued them, in particular people's use of buildings and urban space, and they wished to become psychogeographers, who desired an understanding of 'the precise effects of the geographical setting, consciously managed or not, acting directly on the mood and behavior of the individual' (Debord 1958: 68). Their methodology was that of the dérive, an urban drift, through which they explored places psychogeographically. Psychogeography can be described as 'the hidden landscape of atmospheres, histories, actions and characters which change environments. The lost social ley lines which make up the unconscious cultural contours of places' (Baker 2003: 331).

It focuses on the 'hidden, forgotten and obscure' within 'the settings and practices of the streets, in their fragments, everyday materials and detritus' (Pinder 2005: 389). As a practice that explores experiences of place through ambience, atmosphere and mood, psychogeography retains a sense of the marginal and was an inexact science that dealt with imprecise data, and attempted to combine both subjective and objective modes of study (Sadler 1998: 77). The preferred method for gathering data was the practice of the dérive. With its roots in the nineteenth century flâneur and later forms of Surrealist strolling, the dérive, or drift, was seen as a form of spatial and conceptual investigation of the city. Much of the Situationists' approach to the dérive borrows directly from the flâneur's approach, with participants allowing themselves 'to be drawn by the attractions of the terrain and the encounters they find there' (Debord 1958: 65). However, whilst this may sound as if chance plays a great part in what one might follow and where one might end up, the Situationists believed that the psychogeographic atmospheres and moods at play within the city would actively encourage entry or exit to and from certain areas (Debord 1958: 65). The results of these drifts were primarily two-fold – a series of psychogeographic reports and maps that focused predominantly on areas within Paris. In tandem with the new way of surveying the city, the maps were seeking a corresponding new visual way to represent these findings and were created using the practice of détournement. Détournement is an artistic and often political technique where works of art or mass culture are reworked or placed in different surroundings, so that the original piece is called into question or a new meaning arises. In this context, the psychogeographic maps of Paris were created using existing maps of the city, with places that the Situationists felt were no longer worth visiting literally cut out of the reworked map as they had been spoiled by capitalism and bureaucracy (Sadler 1998: 61). The maps were also an attempt to represent deeper, more complex issues within everyday life, the ambience of different areas and an urban navigational system that was independent of the dominant patterns of circulation within Paris (Sadler 1998: 88). Massey suggests that the maps constructed by the Situationists 'seek to disorient, to defamiliarise, to provoke a view from an unaccustomed angle' and expose the 'incoherences and fragmentations of the spatial itself' (Massey 2005: 109).

Psychogeography has had something of a resurgence in the last two decades, with writer Iain Sinclair bringing it to many people's consciousness with books such as *Lights Out For The Territory* (2003) and *The Last London* (2017), and the affective dimensions of place becoming ever more central to contemporary geographic thinking. For Sinclair: 'Walking is the best way to explore and exploit the city . . . Drifting purposefully is the recommended mode, trampling asphalted earth in alert reverie, allowing the fiction of an underlying pattern to reveal itself' (Sinclair 2003: 4).

Walking as an urban text

De Certeau (1984) writes about space and place using the metaphor of language, and in particular, speaking and writing, and his approach to the study of everyday life and place was an ethnographic one. He describes his ethnographic findings as an urban text, created by 'ordinary practitioners of the city', those who 'live down below', who compose this urban text as they walk within the city. De Certeau (1984: 93–97) compares the act of walking to speaking – 'it is a spatial acting-out of place (just as speech is an acoustic acting-out of language)' and he compares the twists and turns of pedestrians to 'turns of phrase' – it is a 'rhetoric of walking' (de Certeau 1984: 100). It is this everyday act of walking, or 'chorus of idle footsteps' that 'give shape' to place and 'weave places together' and de Certeau sees these 'intertwined paths' as being the 'real system' that 'makes up the city' (de Certeau 1984: 97). The urban text that pedestrians compose could be described as a narrative or story, and de Certeau states that 'stories about places are makeshift things . . . composed with the world's debris' and that things 'extra and other . . . insert themselves into the imposed order' (de Certeau 1984: 107). Seemingly anticipating postmodern geographic definitions of place, he suggests that: 'The surface of this order is elsewhere punched and torn open by ellipses, drifts and leaks of meaning: it is a sieve order . . . articulated by lacunae' (de Certeau 1984: 107). A comparison with Massey reveals striking similarities: 'If you really were to take a slice through time it would be full of holes, of disconnections, of tentative half-formed first encounters, littered with a myriad of loose ends . . . A discourse of closure it ain't' (Massey 1997: 222).

The position and rhythm of walking

That walking is now regularly discussed within the context of an ethnographic practice and methodology could be the result of an interest in mobility, movement, flow and place within geography and a range of other disciplines (Pink et al 2010: 3). It offers an embodied engagement with the environment, and rather than emphasising the researcher's position of being an outsider entering *into* the field, positions them as 'walking *with* – where "with" implies not a face-to-face confrontation, but heading the same way, sharing the same vistas' (Lee & Ingold 2006: 67: italics in original). One enters the flow of place and becomes a part of it. Solnit describes the practice of walking as:

> the intentional act closest to the unwilled rhythms of the body, to breathing and the beating of the heart. It strikes a delicate balance between working and idling, being and doing. It is a bodily labour that produces nothing but thoughts, experiences, arrivals.
>
> (Solnit 2001: 5)

She goes on to suggest that it is the rhythm of walking that 'generates a rhythm of thinking' (Solnit 2001: 5) and indeed, Nietzsche observed that all truly great thoughts are conceived while walking (Burkeman 2010). Many of the quotes are perhaps in danger of romanticising the idea of walking, and allude to previous incarnations of the privileged role of the flâneur who never had to increase his pace (for they were mostly men) beyond a stroll. For some, walking is a means of getting from A to B, a necessity due to financial constraints or limited transport options. Therefore, ideas of 'idling' and 'being' are perhaps far from the thoughts of walkers who have no other choice. However, the familiarity of a regular walk actually offers the walker the opportunity to switch off as the familiar terrain means they need to pay less attention. In turn, this creates space and time within which the walker moves back and forth between paying attention and switching off, enabling them to make connections between small shifts and changes in their everyday environment and their 'more abstract conceptual thought processes' (Ramsden 2017: 58; see also Edensor 2010: 70). This has distinct similarities to Clifford's (1998: 34) ideas of shuttling back and forth between being inside and outside of place and once again alludes to the idea of making the familiar strange. It utilises the slow pace and simplicity of walking to enable one to see what previously one might have never noticed when passing through on a bus or in a car, or by simply repeating the same walk multiple times. Indeed, as Pinder suggests, urban interventions by walking artists offer a method for developing a critical approach to the geography of place as it has the ability to challenge the status quo in terms of the framing and representation of urban space (Pinder 2005: 385).

Walking is also a democratic method through which to understand place. It is free and available to all, although time consuming. Here I am using the idea of 'walking' in an inclusive sense, so rather than just suggest those who are ambulant can 'walk', I include those who might be using self-propelling or electric wheelchairs and mobility scooters. The pace of walking is key, as is its immersive nature. The pace of walking offers the opportunity and the time for reflection (see Edensor 2010: 72, Lee & Ingold 2006: 69). It inevitably slows us down and forces us to engage with our surroundings – our feet are on the pavement, our shoulders may brush other pedestrians or the leaves of trees. We are immersed in the material and affective dimensions of place – it is an embodied practice. Ingold describes this experience as a type 'of sensory participation, a coupling of the movement of one's own awareness to the movement of aspects of the world' (Ingold 2000: 99). When one is walking, one has time not only to look around, but also to think about what one is seeing and develop ideas. There is little else to concentrate on, no wing mirrors to look in and no brakes to apply, just the occasional turn of the head if one is crossing the road or sees something of interest. The sense of simply following one's nose and setting out without a destination in mind can also be useful, along with the idea of this happening over a period of time. Tilley suggests that: 'Understanding place is a gradual process of familiarisation in which description is ultimately the last act . . . It takes time and cannot be hurried . . . One needs to explore first before recording anything' (Tilley 2004: 223).

Reading place

Beyond the researcher's own encounters with people and place, good ethnographic research should also concern itself with other representations of the site of study, for example, fictional and documentary literature, and images and art forms (Pink 2015: 55) as these are embedded in both the representation of place and the making of place. Armin & Thrift (2002: 14) take a similar position specifically in relation to the limitations of the reflexive practice of walking as a tactic for engaging in, and coming to know the city. They suggest that 'a poetic of knowing is not sufficient', and that other, more traditional elements, such as historical guides and photographs, can bring a broader contextual awareness to an attempt to understand and represent the city. In his discussion of top/ography, an approach that is closely related to the geo/graphic one being discussed here, Coles (2014: 516) describes the empirical materials drawn together to develop his narrative of Borough market as drawing from 'the usual sources that otherwise inform qualitative research, such as participant observation, semi-structured interviews, site-writing, archives, and documentary photography'. Archival research, and traditional documentary research that can be undertaken in libraries and utilise newspaper databases, for example, can offer further context to a project. Drawing together a vast body of material like this allows the researcher to interrogate, analyse and understand place from a variety of perspectives.

Ethnographic writing strategies

As we have discussed, ethnographic research inevitably produces inter-subjective, 'inherently partial' truths (Clifford 1986: 7), as different people make sense of events that affect them in different ways and therefore develop their own version of events. These stories or narratives are the way in which the world is 'constructed, understood and acted out', rather than a mirror view of events (Crang & Cook 2007: 14). There are three main ethnographic writing styles: code writing, autoethnographic writing and montage writing. The first is developed from a coding process undertaken with the data. In simplistic terms, this process highlights elements of interest, or patterns within the content. The results can then function as 'heuristic devices' that enable the researcher to think about data in new and different ways (Seidel & Kelle 1995: 54). Autoethnographic writing attempts to use the text to evoke a more personal relationship between the researcher, participants and readers, so here the style and type of narrative is likely to be important. Finally, montage writing is a strategy that juxtaposes and combines different elements of data to illustrate the fact that various fragmentary parts of the ethnographic material do not necessarily fit together seamlessly to create a 'whole' (Crang & Cook 2007: 151). This third style evokes the construction of Benjamin's (1999) *Arcades Project* and Massey's conception of place. In the *Arcades Project*, the structuring of the collection does not follow a traditional linear narrative, rather it exists as a montage of writings, juxtaposed in a 'constellation' that brings the past together with the present, producing a dialectical image. This image, or combination of elements 'produces a spark' that allows for recognition, for legibility, for communication and critique (Highmore 2002: 71).

Such a polyvocal text is also not simply a way of avoiding the responsibility of authorship, but is a means of generating 'perspectival reality' (Tyler 1986: 125). Soja (1989: 1–2) subscribes to this move away from traditional linear narratives and encourages breaking out 'from the temporal prison house of language'. However, he uses Borges' tale of *The Aleph* (1999: 274) to show the potential impossibility of this task. *The Aleph* describes a place where all places on earth can be seen at once, simultaneously, but because of this simultaneity, it cannot be translated into text due to the sequential nature of language. However, as Chapters 5 and 6 will discuss, this issue can be addressed beyond the confines of the traditional pages of an academic book or journal if text and image are brought into play with physical aspects of format and layout.

Such experimental writing does have its critics, who judge its complexities as existing for their own sake and obscuring the real issues of the text (Crang & Cook 2007: 154). However, in other fields, such as art, film and music, montage has been an accepted form of practice since the 1920s. As we have seen, artists such as the Surrealists intentionally used the technique of montage to bring disparate items together in an attempt to disrupt and surprise and thus make the everyday unfamiliar and the ordinary strange. Highmore (2002: 93–95) suggests montage is the most appropriate vehicle for the representation of everyday life and place for three reasons. First, the practice of montage enables diverse accounts and perspectives to be set alongside each other, thus emphasising their differences and offering an opportunity for a critical reading of place. Second, montage also enables these differences to be positioned as simultaneous, illustrating that, in terms of social, economic and cultural factors, development is rarely even. Third, montage resists the collapsing of these diverse elements into a resolved narrative, rather, it presents 'a critical totality of fragments' that position the 'world as a network of uneven, conflicting, unassimilable but relating elements' (Highmore 2002: 95).

Marcus suggests that ethnographers could develop their writing through the use of a 'cinematic imagination' and proposes that writing is effectively a form of architecture that can create 'spaces for imagination' (Marcus 1994: 45) – much like Frascara's (2006: xiv) 'spaces of interpretation'. Rather than use a filmic metaphor, Pile (2002: 204) has suggested that film is simply the best media for capturing multiple experiences and the 'flow of life of the city' as it has the ability to move through time in a non-linear fashion, cut between places, and zoom in or pan out. However, the idea of a sequence of time-spaces does not solely reside with film, but also with books (Carrion 2001; Mau & Mermoz 2004: 33; Hochuli 1996: 35). This frames the opportunity to explore Marcus' ideas about cinematic imagination, and the writing and designing of an ethnographic montage approach through the structure, rhythm, pace and narrative within the temporal form of the book.

Since the advent of postmodernism and poststructuralism, ethnographers have employed a variety of alternative textual formats, such as ethno-drama and poetry, in order to construct 'more open and "messy" texts' (Hammersley & Atkinson 2007: 204). Messy texts are those that explore multiple, non-linear narratives and seek to 'portray the contradiction and truth of human experience' (Guba & Lincoln 2005: 211). However, because of the prejudice toward the written word within the social sciences (Dwyer & Davies 2010: 92, Hammersley & Atkinson 2007: 148), there has been little sustained experimentation with the possibilities of the visual. Scheider & Wright (2006: 4) go as far to suggest that the discipline of anthropology suffers from 'iconophobia', suggesting that there is potential for 'productive dialogue' and 'fertile collaboration' between contemporary anthropology and art that could 'encourage border crossings' and enable the development of new 'strategies of practice' (2006: 1). Pink (2009: 137) has also suggested that 'no conventions for visual-textual sensory evocation in ethnographic texts' have been established and that we need to 'understand the potential of the text that combines still images and written words to represent/describe and comment on the multi-sensory experience of walking and the affective dimensions of this' (Pink et al. 2010: 5).

Writing: a method of inquiry

Previously, we have discussed various types of writing, such as montage writing and creative writing, from a representational perspective. However, writing can also be considered as a 'method of inquiry' itself, rather than just a 'mode of telling' (Richardson & Adams St. Pierre 2005: 923). 'Writing is also a way of 'knowing' – a method of discovery and analysis. By writing in different ways, we discover new aspects of our topic and our relationship to it. Form and content are inseparable' (Richardson & Adams St. Pierre 2005: 923). Richardson is an ethnographer and one of her motivations for reframing the practice of writing as 'creative analytical processes' (CAP) is her sense that much qualitative research, although perhaps rooted in a topic that is extremely interesting, is merely scanned as it often ends up as a text that is not engaging for the reader. There are again parallels here with Pink's concerns and both Mermoz's and Ingold's discussions of typography, print and the place of the page.

Perec (1997: 13) states that 'this is how space begins, with signs traced on a blank page', and Pearson & Shanks (2001: 132) also liken the process of walking through a landscape to the construction of a text.

> It begins with a sheet of whiteness, at once both a page and landscape, a field for action. As a page it awaits our mark. Its whiteness challenges us to begin . . . First nothing, then a few signs which orientate us, and those who follow us, a rudimentary map. So writing plots a journey.
>
> The journey writing plots is inevitably a slow one, but this lack of speed – much like walking – is productive and forces us to . . . perceive actively, to make connections, to articulate thoughts and feelings which would otherwise remain at pre-reflective . . . level of consciousness.
>
> (Tilley 2004: 223–224)

Criteria for evaluation

As we have seen previously, questions remain as to how we might assess research outputs that deviate from the academic norm. Much of Richardson & Adams St. Pierre's approach is informed by the changes in ethnographic writing brought about by the types of postmodern and poststructuralist thinking referred to previously – for example, the sense that there is 'no single way – much less one "right" way – of staging a text' (Richardson & Adams St. Pierre 2005: 936). This sense that within postmodern terms there is no fixed point that can be triangulated – the traditional method of validation within the social sciences – leaves the validity of such experimental practices open to question in some quarters. However, Richardson & Adams St. Pierre suggest that a more appropriate term for the validation of such postmodern texts is that of 'crystallisation'. As an infinitely variable, multi-faceted shape that both reflects and refracts, a crystal offers structure, but one that is perhaps more appropriate than the fixed two dimensional triangle.

> What we see depends on our angle of response . . . Crystallisation provides us with a deepened, complex, thoroughly partial understanding of the topic. Paradoxically we know more and doubt what we know. Ingeniously we know there is always more to know.
>
> (Richardson & Adams St. Pierre 2005: 934)

Richardson & Adams St. Pierre offer a range of criteria that one can use as 'lenses' to see a 'social science art form' and assist with the creation of 'vital' texts that are 'attended to' and that 'make a difference' (Richardson & Adams St. Pierre 2005: 923). These criteria are as follows: substantive contribution, aesthetic merit, reflexivity, impact and expression of a reality. Each of the criteria has specific questions that can be asked of the text, for example, does this piece contribute to our understanding of social life; is the piece successful from an aesthetic perspective, is it satisfying, complex and not boring; how has the author's subjectivity been both a producer and a product of this text; does the piece affect me emotionally or intellectually, does it generate new questions; and does the text embody a fleshed out sense of a lived experience and is it credible (Richardson & Adams St. Pierre 2005: 937)? Their criteria draw on key elements of propositions within all phases of a geo/graphic approach – from 'data' collection, to analysis, idea generation and visual execution. Conversely, many criteria discussed in relation to graphic design tend to either focus on the specifics of form, or make broad references to clear communication. For example, Chen et al. (2003) suggest ten criteria for the assessment of a piece of graphic design: balance, contrast, repetition, gradation, symmetry, harmony, proportion, rhythm, simplicity and unity. Attention to the form of the work is clearly important, indeed Richardson & Adams St. Pierre refer to it, but it is not the sole concern of this research.

However, as we have discussed previously, graphic design has the potential to engage at a level beyond the superficial and aesthetic, and in this framing of writing as a method of inquiry, there are clear parallels with the idea of design as a method of inquiry, or 'designerly ways of knowing' (Cross 2007). Both Burdick (1995: no pagination) and Bruinsma (2001: 1) have likened design to writing and if one were to substitute the word 'writing' in Richardson & Adams St. Pierre's (2005: 923) quote with the word 'design', the statement would not only make sense, but would sum up well a geo/graphic approach to everyday life and place.

Design: a method of inquiry

In the past few years, 'design thinking' has become a well-used term, both within and outside of design circles – perhaps most notably in business. Most closely associated with global design company IDEO and Stanford University, it describes an approach to design that is human centred and more often than not, participatory, collaborative and multi-disciplinary. The approach is described as a series of five steps, which essentially break down the 'design process' into discrete elements. This is useful, particularly for non-designers, but it shouldn't be taken as a statement as to the linearity of the process. Like any kind of research, projects rarely progress seamlessly from one phase to another, and in reality often shift back and forth between phases. In design thinking the terms used to describe the elements of the process can vary, but the goal or action of the step remains the same. First, the designer needs to discover/empathise, second they need to interpret/define, and third, ideate. This is followed by experimentation/prototyping and finally, evolution/testing. Whilst at first glance this would seem to position the research and analysis centred phase of any project within the first two steps, in reality, analysis happens throughout the process, and like writing, also takes place through the crafting of potential solutions. Thus the hands on 'designerly' part of the process – from ideation to evolution/testing – is not only key to crafting a representation that engages the reader, but it is also key to further understanding the facets of everyday life and place under study.

In design, much analysis takes place through the process of reflection. Reflection occurs both as the work is being undertaken (reflection-in-action), and afterwards (reflection-on-action). Reflection-in-action is essentially improvisational, with judgements exercised during the ongoing practice; reflection-on-action retrospectively exercises the skills of analysis and evaluation. Both these types of reflection bring to bear the previous experience and knowledge of the designer – what Schön terms, 'knowing in action' – and enable this to be used in a critical way, rather than an unquestioned, intuitive one (Schön 1987: 28). Often, intuitive decisions taken in practice remain unarticulated, and this has led to a separation between thought and action, and research and practice (Gray & Malins 2004: 22; see also Yee 2007: 8–9). Schön (1991: 308) also suggests that reflection-in-action is what 'recasts this relationship' and addresses that divide, building a relationship between theory and practice. Many readers of this book will not have a design background, so articulating this critical, reflective approach is crucial if these seemingly intuitive moves are to be understood. An advantage of a practice-led approach is that it is a means of 'generating new data through real experiential activity' (Gray & Malins 2004: 105).

However, this approach does have some disadvantages and unless the practice is framed within a clear and transparent methodological approach it can be 'open to criticisms of indulgence and over-subjectivity' (Gray & Malins 2004: 105). Whilst one can obviously never be purely objective, if one's reflection is solely informed by one's own design practice, then this could lead to a perception that work becomes repetitive – a product of a formulaic application of previously experienced strategies that leads to a series of circular journeys from each new problem to each new solution. The onus is therefore on the researcher to ensure rigour in the research. In this case, the reflective approach brings to bear not only my previous experience as a practitioner, but also the wider understanding of the broad, multi-method approach and the relevant theoretical and conceptual positions linked to both the geo/graphic research and practice.

Much like the design process, a geo/graphic methodology not only offers the opportunity to pursue clear developments and directions in the visual work, it also offers the opportunity to capitalise on tangential thoughts and 'mistakes' through more exploratory work. Schön (1991: 141) states that 'reflection-in-action necessarily involves experiment', and Mau (Maclear & Testa 2000: 88) also urges fellow designers to: 'Love your experiments (as you would an ugly child) . . . Exploit the liberty in casting your work as beautiful experiments, iterations, attempts, trials, and errors. Take the long view and allow yourself the fun of failure every day'. This and many others of Mau's (Maclear & Testa 2000: 88–91) strategies explore the possibilities of leaving the well-trodden route and exploring hidden territories in order to find new answers. However, the use of Schön's reflective approach here is crucial, ensuring the practice ultimately progresses through these 'deviations' or 'failures', rather than lose its way. A 'failure' in research terms often provides more information as to why things may not have worked than a final 'successful' piece. In design, this process has been likened to that of a conversation or argument between the work and the designer – the designer takes a decision which in turn changes the direction of the work; this new 'situation' 'talks back'; the designer responds, and so on and so forth (Glanville 1999: 89; Rittel 1987: 2; Schön 1991: 78). Many of these conversations are generated through the prototyping of ideas, and prototyping can be seen as 'an activity to concretise thoughts and make them visible' (Mattelmäki & Matthews 2009: 6). Prototypes, 'failures', and the 'conversations' about them, are therefore vital to developing a clear understanding of the research process, and in this case, developing the geo/graphic design process. Reflection is therefore a generative process, not simply one that is conducted 'in the rear view mirror'. Reflective conversations are often initially held with oneself, or articulated privately within the pages of a sketchbook or journal. However, at some point, these are usually discussed, articulated and tested with others. In interdisciplinary work these conversations may be with collaborators from an external discipline, and in participatory work they may be with research participants. By making these private 'conversations' public one is essentially testing the thinking one has applied to a particular issue and the language used to describe it. McIntosh (2010: 47) suggests that it is this move 'from private to public' that can be enlightening; that the knowing moves from unconscious to conscious forms; and that 'it is through finding the right words that understanding occurs'.

Noticing, collecting and thinking

The analysis of qualitative data seeks to develop explanation, understanding or interpretation of the facets of everyday life and place under study. Seidel's (1998) qualitative data analysis process of 'noticing, collecting, and thinking' offers a simple process for this and reflects and complements the design process, in that it is not linear but 'iterative, progressive, recursive and holographic' (Seidel 1998: no pagination). Each of these characteristics is present in the design process – it is a cycle or spiral like form that enables the design practice to progress through many iterations; it offers the recursive flexibility to revisit previous ideas in the light of new discoveries; and each step of the process cannot be taken in isolation. As a recursive and holographic process it enables a holistic practice that unites both theory and practice.

Noticing works on two levels; first, it relates to the actual fieldwork – producing a record of the things one has noticed; second, it relates to the coding of data. These codes are, in simplistic terms, a way of highlighting elements of interest, or patterns within the content. They can then function as 'tools to think with' (Coffey & Atkinson 1996: 32) or 'heuristic devices' that enable the researcher to think about the data in new and different ways (Seidel & Kelle 1995: 54). Collecting refers to the sorting of the information that has been recorded and coded. The final step in the process – thinking – is the examination of the materials. This process, if looked at from a design perspective, could be seen as a form of editing. The goals are making sense out of the collection, identifying patterns and relationships in the materials and making discoveries about the subject of your research (Seidel 1998: no pagination). Seidel uses the analogy of sorting pieces of a jigsaw puzzle into relevant sections such as edges, sky and grass to describe the process. However, the picture of the research puzzle is not present at the start, but is the end goal, and Seidel cautions against getting lost in the pieces, or the codes, and losing sight of how all the pieces fit together in the bigger picture (Seidel 1998: no pagination).

Summary

This methodological approach forms the basis of the geo/graphic design process and draws together elements from social science and design research in order to construct an interdisciplinary framework that is not only rigorous, but also offers the potential to act in a reflexive, responsive way, and thus capitalise on emergent issues during each of the phases of a research project. It thus positions a geo/graphic approach as one that is process driven. These multiple theoretical and methodological strands are used to develop a creative intervention into place that results in a deep understanding. A geo/graphic approach is likely to be useful for researchers in a range of disciplines who are looking to develop work that utilises creative methods in capturing and communicating the multi-sensory and ever changing, relational nature of place and everyday life. However, it is not intended as a prescriptive list of theories to be applied, a finite set of methods, or a user's manual. Work of this nature inevitably brings together inter-, cross- and multi-disciplinary perspectives, and therefore researchers from other disciplines are likely to interpret, shape and develop the methodology in ways that offer new and different possibilities. The use of creative methods within cultural geography, and the social sciences more widely, is still developing and a geo/graphic approach adds to the body of work undertaken in this area. It positions design as having the potential to contribute more widely to methodological interventions that, until now, have largely drawn from art and performance in their attempts to deal with the more-than-representational aspects of the everyday.

The following chapters discuss three different research projects, all of which have been developed using a geo/graphic methodology. Each project and chapter focuses on a different aspect and understanding of place in conjunction with different re/presentational strategies. This diversity shows the potential of a geo/graphic approach to engage with a range of different empirical methods and theoretical approaches. Yet, in the diversity of projects an underpinning consistency is evident in the commitment to place as a site of study; a desire to develop an embodied understanding and experience of place; and the aim of developing re/presentations that, rather than fixing place within the realms of language, the page and the book, offer readers interactive spaces of exploration.

Bibliography

Armin, A. & Thrift, N. (2002) *Re-imagining the Urban*. Cambridge: Polity Press.

Baker, P. (2003) 'Secret city: Psychogeography and the end of London' in Kerr, J. & Gibson, A. (eds) *London: From Punk to Blair*. London: Reaktion Books, pp. 323–333.

Bakhtin, M. (1981) *The Dialogic Imagination: Four Essays*. Austin, Texas: University of Texas Press.

Battista, K., LaBelle, B., Penner, B., Pile, S. & Rendell, J. (2005) 'Exploring "an area of outstanding natural beauty": A treasure hunt around King's Cross, London', *cultural geographies*. 12(4), pp. 429–462.

Baudelaire, C. (1995) *The Painter of Modern Life and Other Essays*. London: Phaidon.

Benjamin, W. (1999) *The Arcades Project*. Cambridge, Mass.: The Belknap Press of Harvard University Press.

Borges, J. L. (1999) *Collected Fictions*. London: Penguin Books.

Bruinsma, M. (2001) *Designers are Authors*. Available at: http://maxbruinsma.nl/index1.html?authors.html (Accessed: 30 December 2017).

Burdick, A. (1995) 'Introduction/inscription', *Emigre*. 36, no pagination.

Burkeman, O. (2010) 'This column will change your life: A step in the right direction', *The Guardian*. 24 July. Available at: https://www.theguardian.com/lifeandstyle/2010/jul/24/change-your-life-walk-burkeman (Accessed: 31 December 2017).

Carrion, U. (2001) *The New Art of Making Books*. Nicosia: Aegean Editions.

Chen, Y. T., Cai, D., Huang, H. F. & Kuo, J. (2003) 'An evaluation model for graphic design works', *Sixth Asian Design Conference*. Tsukuba, Japan, 14–17 October Available at: www.idemployee.id.tue.nl/g.w.m.rauterberg/conferences/CD_doNotOpen/ADC/final_paper/287.pdf (Accessed: 30 December 2017).

Clifford, J. (1998) *The Predicament of Culture: Twentieth Century Ethnography, Literature and Art*. Cambridge, Mass.: Harvard University Press.

Clifford, J. (1986) 'Introduction: Partial truths' in Clifford, J. & Marcus, G. E. (eds) *Writing Culture*. Berkeley: University of California.

Cloke, P., Cook, I., Crang, P., Goodwin, M., Painter, J. & Philo, C. (2004) *Practising Human Geography*. London: Sage.

Coffey, A. & Atkinson, P. (1996) *Making Sense of Qualitative Data: Complimentary Research Strategies*. London: Sage.

Coles, B. F. (2014) 'Making the market place: A topography of Borough Market, London', *cultural geographies*. 21(3), pp. 515–523.

Crang, M. & Cook, I. (2007) *Doing Ethnographies*. London: Sage.

Cross, N. (2007) 'From a design science to a design discipline: Understanding designerly ways of knowing and thinking' in Michel, R. (ed.) *Design Research Now*. Basel: Birkhäuser, pp. 41–54.

De Certeau, M. (1984) *The Practice of Everyday Life*. Berkeley: University of California Press.

Debord, G. (1958) 'Theory of the derive and definitions' in Gieseking, J. S., Mangold, W., Katz, C., Low, S. & Saegert, S. (eds) (2014) *The People Place and Space Reader*. New York: Routledge, pp. 65–69.

Denzin, N. K. & Lincoln Y. S. (2005) 'Introduction: The discipline and practice of qualitative research' in Denzin, N. K. & Lincoln Y. S. (eds) 3rd edn *The SAGE Handbook of Qualitative Research*. Thousand Oaks: Sage.

Duncan, J. & Ley, D. (1993) 'Introduction: Representing the place of culture' in Duncan, J. & Ley, D. (eds) *Place/Culture/Representation*. Abingdon: Routledge, pp. 1–24.

Dwyer, C. & Davies, G. (2010) 'Qualitative methods III: Animating archives, artful interventions and online environments', *Progress in Human Geography*. 34(1), pp. 88–97.

Edensor, T. (2010) 'Walking in rhythms: Place, regulation, style and the flow of experience', *Visual Studies*. 25(1), pp. 69–79.

Frascara, J. (2006) *Designing Effective Communications: Creating Contexts for Clarity and Meaning*. New York: Allworth Press.

Glanville, R. (1999) 'Researching design and designing research', *Design Issues*. 15(2), pp. 80–91.

Gray, C. & Malins, J. (2004) *Visualizing Research: A Guide to the Research Process in Art and Design*. Aldershot: Ashgate.

Guba, E. G. & Lincoln, Y. S. (2005) 'Paradigmatic controversies, contradictions, and emerging confluences' in Denzin, N. K. & Lincoln Y. S. (eds) *The SAGE Handbook of Qualitative Research*. Thousand Oaks: Sage, pp. 191–215.

Hammersley, M. & Atkinson, P. (2007) *Ethnography: Principles in Practice*. 3rd edn London: Routledge.

Hawkins, H. (2017) *Creativity*. Abingdon: Routledge.

Highmore, B. (2002) *Everyday Life and Cultural Theory: An Introduction*. London: Routledge.

Hochuli, J. (1996) *Designing Books: Practice and Theory*. London: Hyphen Press.

Home, S. (1996) *What is Situationism? A Reader*. Edinburgh: A K Press.

Ingold, T. (2000) *The Perception of the Environment: Essays in Livelihood: Dwelling and Skill*. London: Routledge.

Kincheloe, J. (2001) 'Describing the bricolage: Conceptualizing a new rigour in qualitative research', *Qualitative Inquiry*. 7(6), pp. 679–692.

Lee, J. & Ingold, T. (2006) 'Fieldwork on Foot' in Coleman, S & Collins, P (eds) *Locating the Field: Space, Place and Context in Anthropology*. Oxford: Berg, pp. 67–86.

Maclear, K. & Testa, B. (eds) *Life Style: Bruce Mau*. London: Phaidon.

Marcus, G. E. (1994) 'The modernist sensibility in recent ethnographic writing and the cinematic metaphor of montage' in Deveraux, L. & Hillman, R. (eds) *Fields of Vision: Essay in Film Studies, Visual Anthropology, and Photography*. Berkeley: University of California Press, pp. 35–55.

Massey, D. (2005) *For Space*. London: Sage.

Massey, D. (1997) 'Spatial disruptions' in Golding, S. (ed.) *The Eight Technologies of Otherness*. London: Routledge, pp. 218–225.

Mattelmäki, T. & Matthews, B. (2009) 'Peeling Apples: Prototyping Design Experiments as Research', *Engaging Artifacts: Nordic Design Research Conference*. Available at: www.nordes.org/opj/index.php/n13/article/view/39 (Accessed: 31 December 2017).

Mau, B. (2000) *Life Style*. London: Phaidon.

Mau, B. & Mermoz, G. (2004) 'Beyond looking: Towards reading. . .', *Baseline*. 43, pp. 33–36.

McIntosh, P. (2010) *Action Research and Reflective Practice: Creative and Visual Methods to Facilitate Reflection and Learning*. Abingdon: Routledge.

Nuvolati, G. (2014) 'The flâneur: A way of walking, exploring and interpreting the city' in Shortell, T. & Brown E. (eds) *Walking in the European City: Quotidian Mobility and Urban Ethnography*. London: Routledge, pp. 21–40.

Pearson, M. & Shanks, M. (2001) *Theatre/Archaeology: Disciplinary Dialogues*. London: Routledge.

Perec, G. (1997) *Species of Spaces*. London: Penguin.

Pétonnet, C. (1982) 'L'Observation flottante. L'exemple d'un cimetière parisien', *L'Homme*. 22(4) pp. 37–47.

Phillips, A. (2005) 'Cultural geographies in practice: walking and looking', *cultural geographies*. 12(4), pp. 507–513.

Pile, S. (2002) '"The problem of London", or how to explore the moods of the city', in Leach, N. (ed.) *The Hieroglyphics of Space: Reading and Experiencing the Modern Metropolis*. London: Routledge, pp. 203–216.

Pinder, D. (2005) 'Arts of urban exploration', *cultural geographies*. 12(4), pp. 383–411.

Pinder, D. (2001) 'Ghostly footsteps: voices, memories and walks in the city', *cultural geographies*. 8(1), pp. 1–19.

Pink, S. (2015) *Doing Sensory Ethnography*. 2nd edn London: Sage.

Pink, S. (2013) *Doing Visual Ethnography*. 3rd edn London: Sage.

Pink, S. (2012) *Situating Everyday Life*. London: Sage.

Pink, S. (2009) *Doing Sensory Ethnography*. 1st edn London: Sage.

Pink, S. (2007), *Doing Visual Ethnography*. 2nd edn London: Sage.

Pink, S., Hubbard, P., O'Neill, M. & Radley, A. (2010) 'Walking across disciplines: From ethnography to arts practice', *Visual Studies*. 25(1), pp. 1–7.

Ramsden, H. (2017) 'A walk around the block: Creating space for everyday encounters' in Shortell, T. & Brown E. (eds) *Walking in the European City: Quotidian Mobility and Urban Ethnography*. London: Routledge, pp. 225–244.

Rendell, J. (2006) *Art and architecture*. London: I B Tauris & Co.

Richardson, L. & Adams St. Pierre, E. (2005) 'Writing: A method of inquiry' in Denzin, N. & Lincoln, Y. (eds) *Handbook of Qualitative Research*. Thousand Oaks, California: Sage, pp. 959–978.

Rittel, H. W. J. (1987) 'The reasoning of designers', *International Congress on Planning and Design Theory*, Boston, Mass., 17–20 August. Available at: https://www.cc.gatech.edu/fac/ellendo/rittel/rittel-reasoning.pdf (Accessed: 31 December 2017).

Sadler, S. (1998) *The Situationist City*. Cambridge, Mass.: The MIT Press.

Scheider, A. & Wright, C. (2006) *Contemporary Art and Anthropology*. Berg: New York.

Schön, D. (1991) *The Reflective Practitioner: How Professionals think in Action*. 2nd edn Aldershot: Ashgate.

Schön, D. (1987) *Educating the Reflective Practitioner*. Jossey-Bass: San Francisco.

Seidel, J. (1998) *Qualitative Data Analysis*. Available at: www.qualisresearch.com/qda_paper.htm (Accessed: 30 December 2017).

Seidel, J. & Kelle, U. (1995) 'Different Functions of Coding in the Analysis of Textual Data' in Kelle, U. (ed.) *Computer-Aided Qualitative Data Analysis: Theory, Methods and Practice*. London: Sage, pp. 52–61.

Shaw, W., DeLyser, D. & Crang, M. (2015) 'Limited by imagination alone: Research methods in cultural geographies', *cultural geographies*. 22(2), pp. 211–215.

Sinclair, I. (2017) *The Last London: True Fictions from an Unreal City*. London: Oneworld Publications.

Sinclair, I. (2003) *Lights out for the Territory*. London: Penguin.

Soja, E. (1989) *Postmodern Geographies: The Reassertion of Space in Critical Social Theory*. London: Verso.

Solnit, R. (2001) *Wanderlust: A History of Walking*. London: Verso.

Tester, K. (2015) 'Introduction' in Tester, K. (ed.) *The Flâneur*. Abingdon: Routledge.

Tilley, C. (2004) *The materiality of stone: Explorations in landscape phenomenology*. Oxford: Berg.

Tyler, S. (1986) 'Post-modern ethnography: From document of the occult to occult document' in Clifford, J. & Marcus, G. E. (eds) *Writing Culture*. Berkeley: University of California, pp. 122–140.

Wenger, E (1998) *Communities of Practice: Learning, Meaning and Identity*. Cambridge: Cambridge University Press.

Wolfinger, N. H. (2002) 'On writing fieldnotes: Collection strategies and background expectancies', *Qualitative Research*. 2(1), pp. 85–95.

Yee, J. (2007) 'Connecting practice to research (and back again to practice): Making the leap from design practice to design research', *Journal of Design Principles and Practices*. 1(1), pp. 81–90.

5 Non-linear narratives of food, belonging and multi-culturalism

Food Miles – a culinary journey from Kingsland Road to Stamford Hill

Introduction

Food is one way in which diverse diasporic communities all over the world maintain links with a sense of home and distinguish themselves from other communities within the city they have relocated to. The London borough of Hackney brings together a great number of such communities, some who are well-established and relatively large, others who have arrived more recently, and are much smaller. Such super-diversity (Vertovec 2007) is readily apparent in the presence of a multitude of different ethnic grocery stores, cafés and restaurants throughout the borough. These establishments offer a powerful, multi-sensory way of experiencing place as Zawieja captures in her description of Mare Street, Hackney:

> Mare Street cuts through East London like a diagram. In a straight line from north to south one smell follows another. Fried chicken, sweet potatoes, saltfish pie, banana cake. Spring rolls, lemon grass, soy sauce, fish sauce. Cumin, ginger, dal, coriander, cinnamon, saffron. Lamb, yoghurt, sesame, mint, hummus, thyme
>
> (Zawieja 2009: 141)

Along the old Roman Road in Hackney – the A10 – cafés, restaurants and grocery shops from the more established communities such as the Vietnamese, Turkish, African-Caribbean and Jewish form evident territories. Throughout the rest of the three-mile stretch of road, a range of other ethnicities are represented, including Ethiopian, Brazilian and Polish. On the surface, these borders seem porous, with pedestrians moving into and through these territories at will, seemingly 'rubbing along' (Watson 2006; 2009) with their fellow residents. However, Wessendorf suggests that whilst in public such spaces of superdiversity are characterised by an 'ethos of mixing' (2013a) and a 'civility towards diversity' (2013b), private spaces remain bounded. Using an ethnographic approach, this project explores Hackney's multi-cultural food landscape in the context of how ethnicity is formed, framed and encountered in place.

The iconic image of the map that is perhaps most closely associated with geography ultimately flattens place, positioning the viewer external to it and restricting any reflection of temporality. Unlike the map, the book is essentially a four dimensional space – a 'space-time sequence' (Carrion 2001) – and therefore, working with the material form and physical space of the book, one is able to reflect the ongoing and relational nature of place. In this particular context, the book offers both a local and global narrative of food and migration that co-exist within Hackney. This multi-linear approach to the narrative via the unusual navigation and numbering system, along with the design interventions made within the book, offer the reader an inter-active experience. This enables them to remake place on a third level (Pink 2015: 125), and shows the potential inherent within text and the printed form to generate ethnographic material that engages the reader.

Hackney

This project was developed as part of a series of geo/graphic projects undertaken within the London borough of Hackney. The borough was chosen for its evident diversity and complexity and its contrasting juxtapositions – though this could equally be said of many boroughs within London and many areas throughout the world. Situated in the east of London, Hackney has developed something of a national media presence during the past 20 years. In 2006, Hackney was described as the nation's worst place to live in the Channel 4 programme *The Best and Worst Places to Live in Britain*, which cited pollution and crime levels as key to its decision. In 2001, Upper Clapton Road, to the west of the borough, was labelled 'the murder mile' by the media due to the eight fatal shootings reported in the previous two years. Obviously each one of those sweeping statements over-simplifies a complex narrative and presents a potentially skewed understanding of place. Writers such as Iain Sinclair, particularly within the pages of *Hackney, That Rose Red Empire* (2009), have attempted to offer a broader picture of the borough, one that accepts what some might see as 'failings' and celebrate these as part of the particular 'spirit' of place. In terms of 'place image' (Shields 1991: 47) this type of negative publicity might be considered as sounding the death knell for an area. Yet, as with many areas in large global cities like London, other financial and housing pressures create the kind of forces that combine to produce different results. These days Hackney is now fêted as one of the capital's coolest places to live, with rents and property prices having seen vast increases and once derelict, unloved spaces such as Broadway Market redeveloped into thriving commercial spaces. Hackney is now well into the throes of gentrification and for some has gone past the 'tipping point', with artists priced out of the area, industrial build-ings that once housed studio spaces being sold off for luxury flats and mainstream brands like Aesop and Prêt a Manger moving in alongside independent traders. However, whilst the hipsters, smashed avocado and gentrification have become one of the dominant narratives of place, Hackney remains much more diverse than this.

Hackney is home to around 270,000 residents who, as you might expect, rep-resent a diverse multi-cultural mix of different races and ethnicities. The largest ethnic groups within Hackney are as follows: 36.2% are white British; 23.1% are black African, Caribbean, or black British; 18.5% other white; 6.4% are mixed/multiple ethnic groups; and 6.4% are Indian, Pakistani or Bangladeshi or British Asian. Hackney therefore has a smaller white British population than either London (44.9%) or England as a whole (79.8%). Conversely, it has a larger black population than either London (13.3%) or England (3.4%) (Policy and Partnerships Team 2017: 1). Hackney has traditionally been a place that has seen waves of migrants arrive from across the world, going as far back as the late 1600s when many Protestant Huguenots fled France following religious reformation and approximately 300 families settled in the area. Situated in the East End, and therefore close to the West India and East India docks, opened in the early 1800s, Hackney is geographically positioned as a place of arrivals and thus the borough has seen the formation of relatively large ethnic communities. For example, the Charedi orthodox Jewish community, predominantly based in and around Stamford Hill to the north of the borough began arriving in the 1920s but expanded greatly prior to, and during, the second world war. Hackney also has a large Turkish community; Commonwealth citizens arrived in the 1930s and economic and political migrants in the 1970s and 1980s. There is also a well-established Kurdish speaking community, the majority of whom fled persecution in Turkey, Iran and Iraq in the late 1980s and early 1990s. The sizeable Caribbean population in Hackney is diverse, having predominantly grown since the 1960s and 1970s. The Vietnamese community was established in the mid 1970s, and large communities from countries such as Nigeria, Ghana, Congo, Senegal, Sierra Leone and Uganda are also present, most of whom migrated in the 1960s and 1980s. Hackney is also home to a large South Asian community, who began arriving in the 1950s and 1960s, the largest groups of which were originally from India and East Pakistan (now Bangladesh). Finally, a relatively large Eastern European community has developed since 2004, most notably from Poland (*History of Hackney's Diverse Communities*, 2016).

Many of these residents migrated to London due to political upheaval, unrest or economic hardship, and as part of the process of settling in London they negotiate their relationship with shifting perceptions of 'home' and belonging. For diasporic communities, food plays a major role in this negotiation, offering a link to homelands that have been left behind and providing a community hub within unfamiliar surroundings.

Food, memory and belonging: creating a home away from home

For those displaced or migrating, a sense of home, and therefore a sense of belonging, is continually 'readapting and readjusting' to new and changing realities (Marcu 2014: 327). In this context, ideas of home and a sense of belonging are developed through the interrelationship between the material site of home in the present and the remembered and reimagined site of home in the past. Food plays an important role in this, and when settling in new countries, as part of developing an infrastructure within that which they find on arrival, diasporic communities inevitably seek to set up grocery shops, cafés and restaurants that offer a 'taste of home' and a meeting place. These commercial operations effectively become a cultural identity marker, both through marking out their territory in terms of difference from other ethnic or indigenous communities who are already present, and in terms of signifying similarity within their own community.

Much has been written about the power of food in drawing us back to memories of previous people, places and encounters. Often discussed in terms of nostalgia, for expatriate or diasporic communities the sensory memories triggered when eating food from home emphasise their displacement, but provide a momentary return to a time and place when their lives were not fragmented (Holtzman 2006: 367). However, this power isn't confined solely to eating, but includes the process of shopping for ingredients at a specific grocery store or visiting a particular restaurant. The environment of each evokes a range of multisensory 'cultural mnemonics' (Mankekar 2002: 86), whether that be through the familiar décor and packaging, or particular sounds and smells. However, the narratives of the past triggered by these mnemonics will be personal and different for each visitor. Constructing the past is always a process of selectively remembering and forgetting that is informed by one's particular perspective and aspects such as class, gender and age. This subjectivity notwithstanding, it is clear that food provides and maintains both temporal and spatial connections to past events and experiences, and it is this multi-sensory potential of food and all it entails, that brings with it a synesthesia-like experience that makes it particularly compelling in terms of memory (Sutton 2010). In this context, food therefore provides tangible connections to 'home' for those who have left, but for future generations it also enables the continuation of tradition and the building of a sense of place about somewhere they may only ever have heard stories about.

Food, multi-culturalism and 'authenticity'

Whilst these grocery stores, restaurants and cafés play a key role in the construction of cultural identity for diasporic communities, they are also businesses and in that sense ultimately cater to a broader population than those sharing their heritage. Commodification of racial 'otherness' has been explored by hooks in her article *Eating the Other: Desire and resistance* (1992). She argues that a range of discourses and practices lead to the celebration of ethnicity as 'a spice, a seasoning that can liven up the dull dish that is mainstream white culture' (hooks 1992: 181). With this comes thoughts of 'the Other' and 'the exotic', and this culinary tourism could perhaps be reframed more sinisterly as culinary colonisation, an expression of power and privilege. Such culinary tourists are often seen as adventurous, desiring 'real' experiences through the participation in different ethnic culinary traditions (Heldke 2003). Their territory is often that of the ethnically diverse inner city, and they have been described by May as the 'new urban flâneurs' (1996: 205) who seek out 'authentic' ethnic experiences in the multi-cultural inner city that border on 'cultural voyeurism' (1996: 206). Such restaurants and their migrant owners or customers, although a product of the literal flows of movement and people through space, somewhat ironically become framed in the mind of such culinary tourists as embodying 'fixity and authenticity' (Karaosmanoglu 2014: 228). Ideas of 'eating the Other' are not without criticism however, and Cook (2008) in particular has questioned the use of a such a limiting binary oppositional thinking. Duruz (2010) has also contested such conventional identity categories and associated attachments to place with her conception of 'floating food' which, with its multiple perspectives, locations and memories, problematises the fixing of identity categories. As Karaosmanoglu notes, there is 'no automatically taken-for-granted, other-eating White personality . . . people have heterogenous biographies and everyday lives' (Karaosmanoglu 2014: 226).

The restaurants and cafés perform this ethnic 'authenticity' by highlighting 'genuine' characteristics and customs from 'home'. Themed restaurants like this turn a service into an experience using 'referential authenticity' which refers to another context and is a deliberate attempt to trigger shared memories and longings. If this referential experience fails to feel 'authentic' it will be judged as fake (Gilmore & Pine 2007: 49–50). 'Authenticity' within the inner city is also sought and acclaimed through a sense of danger or excitement, and here we do return to themes relating to 'the Other'. For example, when researching food bloggers' constructions of Turkish restaurants in Dalston, Hackney, it was clear to Karaosmanoglu that there was a correlation between poverty, dirt and danger and an eating experience that was perceived as vibrant, and 'full of culture' (2014: 228). However, as with any reference to 'authenticity', it is inevitably a social construction – ultimately one person's real is another's fake. Yet, authenticity is what all customers are increasingly seeking (Gilmore & Pine 2007) and therefore business owners also face their own complicated negotiations through this territory as they almost inevitably move from a space that primarily acts as a focal point for a recently arrived diasporic community to a growing commercial business that seeks a broader customer base.

A geo/graphic approach to Hackney's culinary landscape

Hackney is frequently discussed in terms of its multi-cultural population and I was therefore keen to explore a project that might reflect this in some way. However, at the outset of the research I had no clear idea of what this might be. So, my starting point was to explore Hackney on foot using the strategies of 'floating observation' (Petonnet 1982) and 'salience hierarchy' (Wolfinger 2002). Relatively early on within this process I became aware of very visible traces of Hackney's diverse communities through the wide range of ethnic grocery shops, cafés and restaurants. More often than not, these were evident visually both through their signage and use of language (see Figure 5.1), but also often through aural and olfactory means – sounds of languages other than English being spoken, or smells of particular types of food. I identified the three-mile stretch of the A10, which bisects the borough and runs from its southern-most tip to its northern-most, as having a large presence of food related establishments. Given that the A10 follows the path

Figure 5.1 Sign advertising Ethiopian produce

of the old Roman road, it seems appropriate that so many other recently arrived residents would take the same path. In reality, this stretch of the A10 is formed by several different roads that pass through several areas of Hackney: Kingsland Road in Shoreditch and Kingsland High Street in Dalston, along with the eponymously named Stoke Newington High Street and Stamford Hill.

I mapped each establishment into a sketchbook, simply using two parallel lines to define the road and positioning each one in relation to its geographic position in space (see Figure 5.2) On analysis, this revealed over 150 establishments that could be described as 'ethnic', with particularly large numbers of Turkish, Vietnamese, Caribbean and Jewish concerns, most of which represent well-established diasporic communities within Hackney. Following the identification of the different ethnicities present and the numbers of establishments, I then visited 33 of these, ensuring I visited each different ethnicity present and visiting multiple establishments from those communities listed above that have a larger presence in the borough. In cafés and restaurants I ate and drank food on the premises, and in grocery shops I bought food items to take away. The only enforced constraint during this process was my vegetarianism, which obviously meant many dishes, including some that might be considered traditional, such as Turkish tripe soup, were off-limits.

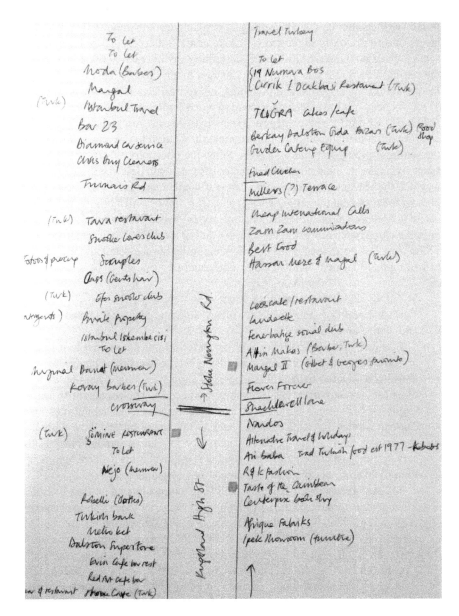

Figure 5.2 Initial mapping of restaurants, cafés and shops

Field notes, field-writing and difference

In ethnography, the engagement with the site of study is traditionally described as fieldwork, with ethnographers recording their observations through the taking of field notes during participant observation and generating further notes that are usually written after leaving the setting. A productive combination of six different layers of description is suggested by Cloke et al. (2004: 201–204): locating the setting; describing the physical space; describing any interaction of people within the setting; describing one's own participation in any interactions; reflecting on the process of the research; and including any self-reflections on the experience. These six headings have been developed in much further detail by Crang & Cook (2007: 51) who offer some suggested questions that the researcher might try to answer within their field notes. For example, in relation to the physical space of the setting they suggest addressing the following questions: what is its size and shape; what are its main physical characteristics; can you describe these in such a way that the reader can picture them; can you supplement notes with drawings or photographs; and did the setting change and, if so, how? Whilst this approach was useful to draft a clear picture that aided my memory when back at my desk, I found myself writing much more reflective pieces during the visit to each establishment. These were less notes than more fully formed, reflexive auto-ethnographic pieces of writing that I termed 'field-writing'. For example:

ŞÖMİNE: 131, Kingsland High Street
Turkish Breakfast: Eggs, cheese, olives, tomato, cucumber, bread, butter, jam, honey and Turkish tea

Şömine is at the crossroads with Shacklewell Lane, so to get to it I have to walk past the end of Ridley Road market. It is another sunny day and Kingsland Road is busy with shoppers. I almost feel that I am abroad as I overhear so many different languages and accents. The restaurant is nearly empty of customers, just a father and daughter at a table behind me. Two Turkish women sit at tables at the back of the place and make the flatbread for the day. I notice that they are open twenty-four hours and that breakfast is served from two in the morning. I wonder if the customers then are getting up for a night shift, or on their way home after an evening out?

My tea arrives in a glass mug with four lumps of sugar. I drop three in. It's a golden colour and the sun streaming through the open windows makes it glow. The breakfast arrives. I like the way the slabs of butter are used to separate the runny honey and jam from each other, and from the cheese. I start with an olive and wonder if there is a correct order in which to eat it all. I know that I will go from savoury to sweet as that is what I am used to.

A steady stream of people are coming in, but most either want something to take away, or are staff. I think the shift must be about to change. The windows are all wide open and as I lean my elbow on the edge, I watch people standing just outside waiting for the lights to change so they can cross the road. As I sit, I wonder if breakfast is the one meal that really defines us as a particular culture or country? I notice that I am the only English looking person in here, unlike last night when the only Turkish people were staff. Is this because it is breakfast, or because Şömine, with its formica tables, is not deemed 'restaurant' enough for non-Turkish customers?

These visits and the food itself in turn triggered a further range of research and writing, much of which enabled a broader, complementary reading of place as prescribed by Pink (2015: 55) and Armin & Thrift (2002: 14). This enabled a greater understanding of some of the cultural signifiers or practices I was encountering, such as 'lucky cats', the chewing of chat, or Shabbat dinner. The various types of writing were also supplemented with visual ethnographic materials, in particular documentary photography.

Before and during these culinary experiences I was well aware of issues surrounding 'cultural colonialism' and 'food adventurers' (Heldke 2003) or 'eating the Other' (hooks 1992), and as a white, western, middle class academic, I might be said to tick many of these boxes. However, much as the artifice of the research situation sometimes brought these feelings of guilt to the fore, one thing enabled me not to assuage guilt, but rather to reposition my thinking and thus my experience. In this context, 'guilt' and the awareness of my multi-faceted position in the world, and thus in the field, is 'transformed as into an effective tool for understanding difference' (Cook et al. 2010: 111). In particular, it was my position as a woman that decentred whiteness as 'a fulcrum of domination' over others (Garner 2007: 175). In several of the establishments I was the only woman eating at that time, and on occasion, the only one unable to speak the language that was being spoken by the majority. In such moments, I was momentarily repositioned as 'the Other' and therefore able to gain a brief insight into this position of difference.

Re/presenting place

Unlike the map which positions the viewer above and outside of the territory, I was looking to develop a re/presentation of place that could better situate the reader within Hackney. However, conceptualising the book as a process, as something that is 'spatialised and relational' enables one to rethink the book in a way that enables a 'dynamic concept of mapping' (Meskimmon 2003: 160). Thus, much like discussions of the map as processually emergent (del Casino & Hanna 2006; Kitchin & Dodge 2007), the book can be positioned as 'a social forum, open to diverse readers/subjects, engaging materially and dialogically with ideas' (Meskimmon 2003: 160). The research for this project was undertaken over a period of months so there was both a spatial and temporal element to it. Therefore the form of the book seemed further appropriate as it unfolds in a 'temporal dimension' (Mau & Mermoz 2004: 33) and is effectively:

> a sequence of spaces. Each of these spaces is perceived at a different moment—a book is also a sequence of moments. . . Written language is a sequence of signs expanding within the space; the reading of which occurs in time. A book is a space-time sequence.
>
> (Carrion 2001: no pagination)

Unlike traditional cartography which eradicates time, a book designer's challenge could therefore be positioned as a 'space-time problem' (Hochuli 1996: 35).

The process of research had generated a range of different types of material with which to re/present place geo/graphically – autoethnographic writing about the experience; writing about an aspect of the experience that I knew little or nothing about previously; further post-experience writing that theorised particular aspects of it; records of brief conversations with staff and owners; and documentary photography. These multiple approaches go some way to developing a version of place that reflects its complexity. However, a strategy for re/presenting such complexity needs to be mindful of not developing a 'confused presentation' (Crang & Cook 2007: 201) and conversely must not systematically reduce the confusion to a level that eradicates ideas of multiplicity, movement and relationality inherent in contemporary definitions of place. A mid-point thus needs to be found that enables the reader to engage with place in a way that avoids a reductive, linear representation. After prototyping a variety of different approaches, I settled on one that differentiated the written texts both typographically and spatially. The field-writing that captured the central experience of eating is positioned vertically, central to the page in what might be considered a traditional book text format. The two types of supplementary texts are set perpendicular to this and run from left to right across the pages, requiring the book to be turned for them to be read – they literally 'stem' from the main text (see Figure 5.3). Each of these texts uses a different typeface – the field-writing texts

Figure 5.3 Supplementary texts 'stemming' from main text

are set in a traditional serif font, the factual texts in a sans serif font and the theorisations about aspects of the experience are set in an italicised serif font. Although the layout retains a complexity in that multiple strands of texts are present, and these are read in different directions, the use of different typefaces creates a system which maintains their distinctly different perspectives.

In analysing these materials, and thus retracing my steps and reimagining place (Pink 2015: 143), I realised that my chronological eating journey was not the only journey present within the research and there were two others evident. First, the local journey that is made along the length of the road that reveals a particular spatiality defined by many of the larger diasporic groups. Second, a global journey that has been undertaken by those groups in moving to London was present. Obviously the majority of books are read in a linear fashion, following the chronology of their page numbers. However, some books – like encyclopaedias or dictionaries, for example – have two navigational approaches, using both page numbers and alphabetised entries. Taking inspiration from this I developed a three-pronged navigational system that enabled the reader to take my journey in chronological eating order; following the local spatial journey from the bottom of Kingsland Road to the top of Stamford Hill; and taking the global journey between London and the ethnic origins of the restaurants, cafés and grocery shops. This is developed through the use of three numbering systems. Traditional consecutive numbers at the foot of each page relate to my own journey, whereas at the outer edge of each page a numerical scale is positioned. On the left hand page this runs from 0 to 3.0 and relates – in miles – to the length of the stretch of the A10 included in the book. The distance from the southern-most tip is recorded at the relevant point using a darker shade of text (see Figure 5.4). On the right hand page, the scale runs from 0 to 5 and relates to the distance from London and the heritage or origin of the café, restaurant or grocery store in thousands of miles. Once again, the distance is recorded at the relevant point in a darker shade of text (see Figure 5.5).

Whilst I wanted to avoid the 'flatness' of a map, I was interested in trying to develop some way of 'opening up' the space of re/presentation to enable the reader to trace a route through the book in the same way that they might search for and plan a route on a map. Unlike a map, books reveal their pages over time, but I was also interested to see how I might use the space of the book in ways above and beyond this that could further develop the potential for multi-linear, interactive readings and experiences. In this context, the book is seen as a four dimensional space, one that has the ability to draw the reader into it by utilising its physical and material form to develop elements that offer interaction and the possibility of discovery. A range of different design interventions have thus been constructed, for example, using a French-fold binding, I have created pages that essentially contain a space within them. With a French-fold, the edges of the A4 sheet of paper are bound into the spine of the book and the sheet then folded, meaning it creates an A5 page that not only has a front and back face as normal, but each of these also have a reverse face within that fold. This means a further space for text or imagery is available to the designer, and for the reader elements

Figure 5.4 Local navigation system

have to be sought out as they are hidden from the normal external page view. Some of these pages also have their leading edge perforated, which means the reader is able to break the French fold open. One section that uses this technique holds a small book within the French-fold, which is only accessible when the perforated edge of the page is torn open (see Figures 5.6 and 5.7). The contents of the inserted book are to some extent tangential to the journey itself, but were triggered by the purchasing of a Polish chocolate bar that is branded as made by E. Wedel, a brand with a long Polish heritage. I discussed the meaning of the brand with a Polish man and it was clear it was something that, for him, represented his country and contained memories of childhood. However, in 1999, Wedel was bought by Cadbury Schweppes. Again, Cadbury's is a brand with a long British heritage and much like Wedel its predominantly purple chocolate

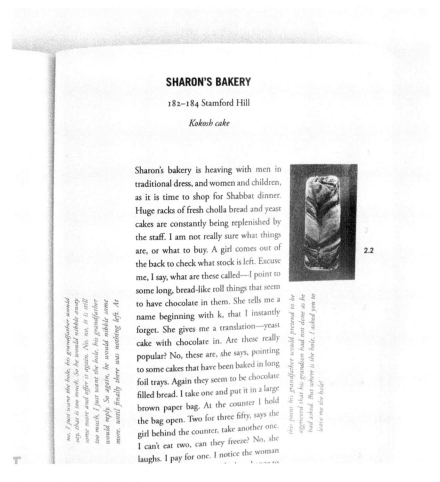

Figure 5.5 Global navigation system

packaging is instantly recognisable on the shelves, and much like Wedel, it too was bought out in 2010 by Kraft. So both brands are now American owned, which begs the question as to whether they can still be considered 'national' products. A further section contained within a French-fold also contains reflections on culinary tourism, food creolisation and 'authenticity'. This section starts on an externally facing page, but can only be read in full if one finds the rest of the text which is within the French-fold. So, much like space, one has to physically explore to find out what might be behind the closed door or round the next corner (see Figure 5.8).

As Richardson & Adams St. Pierre (2005) have observed, much ethnographic writing is in a form that offers little in the way of engagement for the reader.

Figure 5.6 Small insert book visible from above French fold

Thus, I also wanted to use the content and design in such a way as to draw the reader into the stories of these different experiences and to sustain their interaction with the book. Here, along with the previous design interventions, the use of both text and image plays a part in a simpler, more obvious way. The images are used not only as a way to illustrate the discussions surrounding the eating experience and related texts, but from a designerly perspective they are used to enliven the reading experience. So they have been considered in terms of scale, positioning, impact and 'pace'. Pace throughout a multi-page document is important − if the pace remains constant with images and text used in the same

Figure 5.7 Perforations open to reveal small insert book

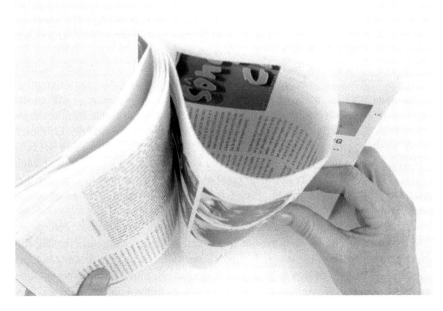

Figure 5.8 Culinary tourism section visible from above French fold

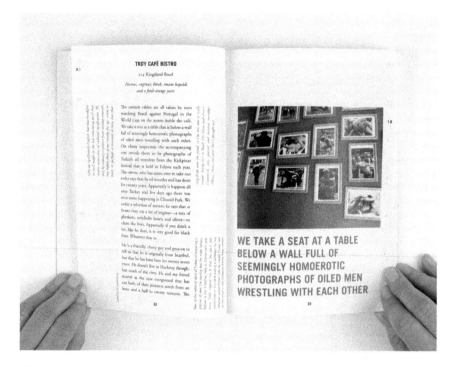

Figure 5.9 Example of pull quote

way on each double page spread, the reader will know what to expect as they turn the page. By varying the size and positioning of imagery, this experience is more varied. Each food experience has been given a simple typographic title in the form of the name of the restaurant set in bold sans serif capitals, the address positioned beneath this set in regular weight serif type and the food eaten positioned beneath this in italicised serif type – this simple hierarchy makes the facts clear. However, to draw the reader into the content, excerpts from the main text are used as 'pull quotes'. For example, 'I'm surprised to find my Ethiopian tea is Yorkshire tea with spices added' and 'We take a seat at a table below a wall full of seemingly homoerotic photographs of oiled men wrestling with each other'. These excerpts are set in larger, bold sans serif capitals in a colour that relates to the photographic imagery within the particular double page spread (see Figure 5.9).

These design interventions conspire to entice the reader to explore place through the material form of the book and endeavour to engage the reader with the content of the text. However, the book is not simply a vehicle to showcase the potential of graphic design to produce an interactive, multi-linear experience of place; work of a geo/graphic nature engages with both form and content, and thus contributes to both re/presentation and understanding.

Discussion

The range of diverse experiences led to a range of varied insights about food, consumption, multi-culturalism and place. For example, as I walked down the A10 and designed the book, I began to think about borders. In terms of contemporary, progressive versions of place, borders are seen as porous and permeable, and Massey's (1994: 153) oft cited description of Kilburn in London illustrates this in a way that resonates closely with Hackney – for example, shops selling goods from all corners of the world and international flight paths overhead. Yet in Hackney, along the various sections of roads between Shoreditch and Stamford Hill, well established communities – such as Vietnamese, Turkish, African Caribbean and Jewish – are represented by food and drink establishments that do seem to demarcate, or perhaps border, particular areas. For example, the Vietnamese restaurants are clustered at the foot of Kingsland Road at the southern-most tip of the borough, and the Jewish shops at Stamford Hill at the northern-most tip. Hackney is often described as a place where multi-culturalism and a multi-ethnic community live relatively happily side by side, one that in contemporary geographic terms might be described as progressive, while in other places such diversity can often be framed within a narrative of division, friction or exclusion. Ang discusses how claiming ones' difference is an important strategy for diasporic communities worldwide and provides 'a powerful sense of transnational belonging and connection with dispersed others of similar historical origins' (Ang 2003: 142). In the context of the borders in Hackney, demarcated by the concentration of specific ethnic establishments, we return to ideas of food as a cultural identity marker, setting the newly arrived residents apart from the indigenous population and other migrant communities, as well as connecting this particular community with each other and with 'home', thousands of miles away. However, for Ang, this border or boundary is a double edged sword, and whilst it provides support, emancipation and solidarity, at the same time it engenders oppression, confinement, and division (2003: 142). These negative consequences lead to a regressive sense of place where the dominant narrative in the UK is more usually one of the previous white, often working class, majority feeling as though 'their' place has changed for the worse.

Yet in this context, despite these oppositional binary framings of inclusion and exclusion, a multitude of encounters occur that inevitably position the city as cosmopolitan and the street as a social space in which those from different backgrounds do interact or, as Watson in her research on markets describes, 'rub along' (2006; 2009). This rubbing along takes place in a variety of ways: 'daily observations of the markets revealed a plethora of fleeting forms of "rubbing along", connecting, lingering and taking pleasure in a shared space for everyone in the market including those who are often marginalised elsewhere' (Watson 2009: 1589). Such interactions have the potential to challenge 'racist discourses and stereotypes of unknown others' (Watson 2009: 1582). However, there is a sense here that whilst this is implicit in the space, it is also at times inadvertent. In a crowded market place or shopping street, one finds oneself jostling for space and literally rubbing along together, brushing against other shoppers as one passes them. This 'rubbing together' aligns with Ang's (2003) ideas of 'togetherness in difference' which she argues is foregrounded through the concept of 'hybridity' and hybridisation, which takes the form of 'exchanges, crossings and mutual entanglements' (2003: 147) – though unlike the exchanges Watson refers to, these are not always harmonious. The use of hybridity in this context is particularly interesting as it manages to simultaneously avoid the binary oppositions of same/different, integrating them both in a way that refuses notions of absolutism. It therefore 'confronts and problematises boundaries, although it does not erase them' and acts as an 'heuristic device for analysing complicated entanglement' (Ang 2003: 149–150).

On the surface of it, there is togetherness-in-difference in Hackney, and this is something that the borough is proud of, that by and large, many cultural groups live side by side without huge problems. However, with such clearly demarcated territories, I wonder just how 'together' these groups really are. Many places proudly display their heritage in different ways – Turkish meat stalls sell tripe; African stalls sell large edible snails; various signs display flags or names of their respective countries; and much shop front signage is written in respective languages (see Figures 5.10 and 5.11). Particular characters that do not exist in the English alphabet pepper the windows, doors and walls, creating a visual language that displays cultural difference via the typography. So whilst there is clearly a 'rubbing along' in terms of daily contact on the street as described by Watson, inside cafes and restaurants there didn't seem to be a consistent experience that might suggest a togetherness-in-difference.

Some establishments were populated almost exclusively by customers who were members of the particular community they were serving. For example, this

Figure 5.10 Shop frontage displaying name of country

Figure 5.11 Sign displaying flags and using native language

was the case in Andu, an Ethiopian café, Tropicalia, a Brazilian café, Centerprise, a Caribbean café, and Sharon's, a Jewish bakery. My experience of each of these places reinforced my sense of difference and this felt as if it variously positioned me as 'the Other' by way of my race, gender, sexuality, and religion (or lack of it). In Andu, for example, I was the only white customer and the only woman. The café clearly acts as something of an Ethiopian community centre and the back room was full of men chewing chat and engaging in lively discussion. They paid no attention to me, had no interest in my being there, they were perhaps indifferent and apart rather than us being together-in-difference. However, for the owner my presence did seem important – he immediately shooed chatting customers who had finished their meal away from a table to free it up for me and proudly told me about his plans for the place. I was brought white sliced bread along with the traditional injera fermented bread, and was served my meal on a plate with a knife and fork rather than on a tray to be mopped up using the bread (see Figure 5.12). I was definitely being treated differently. I suspect the addition of bread and cutlery was to ensure I wasn't troubled by unfamiliar flavours or customs, and to an extent this, and being offered a table immediately, reflects a business owner who cares about his customers and wants to attract more of them in order to achieve commercial success – which may require going beyond success solely within his own ethnic community. Yet it was also evidently a pride in his journey and the food of his country that he was keen to share and this did feel like a moment where we were together-in-difference.

Conversely, other cafés and restaurants were populated by a majority of customers who weren't from the same ethnic background as the establishment – for example, Mangal 2, a Turkish restaurant and New Rice & Spice, an Indian restaurant. In the context of these two restaurants it seemed they were primarily businesses seeking to provide an 'authentic' experience of ethnic cuisine for those who were not necessarily Turkish or Indian, rather than provide a cultural identity marker for their respective communities. It would seem that there is perhaps some kind of spectrum of familiarity at play – in terms of the food for

Figure 5.12 Traditional Ethiopian Injera bread and white sliced bread

those who are not from that community, and in terms of being at home away from home for the ethnic communities. For example, the Turkish community is well established in Stoke Newington and Dalston, and the Asian community well established in the UK generally, so perhaps they have less need for such a clear marker of difference, and other communities have become familiar with their cooking. However, whilst the staff were predominantly Turkish and Indian and we conversed to an extent around the menu, there was less of a sense of togetherness-in-difference here.

One of the underlying issues related to this, particularly in the context of food and drink, is that of 'authenticity'. Previously, competitive leverage in the service industry was gained through the provision of enhanced quality. However, in the contemporary marketplace the management of customer perceptions of 'authenticity' is key (Gilmore & Pine 2007: 3). It has become the new buzzword within marketing, and is sought in terms of both experience and product by consumers who increasingly base their purchasing decisions on 'how real or fake they perceive various offerings' (Gilmore and Pine 2007: xi). In the context of the ethnic restaurants in Hackney, many of these issues around 'authenticity' are entangled and difficult to separate, not just in theory, but also during the practice and experience of the eating itself. Are the lucky cats in the windows or on the counters of the Vietnamese shops and restaurants there because that is what one expects, or because the owner wants to ensure the protection and continued success of their business (see Figure 5.13)? Are the shrines with the paper money there for show

Figure 5.13 Lucky cat in grocery shop window

or for honouring the dead? Is it what I and other customers expect, or is it actually a private act being played out in a public place? Are the women in Şömine just making the bread for the day, or are they 'performing' for the customers and at the same time signifying that the bread is homemade and thus 'real'? Are the television sets part of the experience too, and if so, are they for bored customers or idle waiting staff?

Gilmore & Pine (2007: 89) assert that 'all human enterprise is ontologically fake', but 'output from that enterprise can be phenomenologically real'. In other words, such restaurant experiences can be perceived as authentic by the customer. 'Authenticity' is therefore a social construction and it is the interaction between the customer and all facets of the experience that creates this belief (Gilmore & Pine 2007: 109–110). One can therefore only comment on 'authenticity' from one's own perspective. In my experience, the television in the Hanoi Café was positioned in the corridor between the bar and kitchen, not a space a customer would pass through. There was one chair in front of it. This was there for the waitress during the slower hours of the afternoon; it was not for me. Conversely the television in Tropicalia is positioned so that both customers and staff can watch; it seems to provide information and entertainment, often via Brazilian channels. Timur Öğüt (2008: 77) suggests that the bread making women in some Turkish restaurants are not providing a designed and preformed act of authenticity, as they are dressed in ordinary clothes, oblivious to customers and working as if cooking at home. I wonder if this is always the case though, as some of the Turkish restaurants place the women directly in the window fronting onto the main street. This doesn't seem an inconsequential positioning and seems to be a thought through decision that assists in communicating not only a sense of 'authenticity', but also ideas of 'tradition', 'home/hand-made', 'provenance' and 'freshness' – words that are often used in conjunction with the concept of 'authenticity'.

Many ethnic restaurants do have a performative aspect that is overtly played out through elements such as the décor and various cooking practices. In many of the Turkish restaurants, for example, it seems that this performance is primarily for customers from a different ethnic background to experience a 'traditional' or 'authentic' taste of Turkish hospitality. Instead of reinforcing a togetherness, this once again reinforces difference, and whilst this might be positive in terms of enabling an acceptance or even celebration of that difference by non-Turkish customers – therefore moving towards Ang's notion of hybridity, it possibly does so at the expense of attracting local Turkish customers and therefore simultaneously moves away from ideas of togetherness and hybridity. It would seem that at both ends of the spectrum – those establishments that primarily focused on customers from their respective ethnic communities, and those that didn't – there are barriers to 'togetherness'. It would also seem that some of these may be built by a business model that seeks to develop these establishments beyond their original aspirations of providing a taste of home, to a broader commercial concern. In terms of branding and marketing, received wisdom is that the more clearly one's target audience is identified, the more successful one will be at positioning one's offering and attracting customers. Therefore, this idea of 'together-in-difference, which implies an audience that is made up of multiple types of customers and some kind of 'middle ground' between the two extremes discussed previously, is less likely to work well in a business context. As Wessendorf (2013a) found in her research on Hackney, while many people mix and interact in external public spaces, they usually return home separately and private spaces are rarely realms of cultural difference. I would suggest that something similar could also be said for many of the various ethnic restaurants, cafés and food stores.

Unlike Ang's togetherness-in-difference and Watson's 'rubbing along', multi-cultural community cohesion – or the lack of it – has often been discussed in the context of 'parallel lives' and 'layers of separation' (Cantle 2008). In this context, the two main groups under discussion in the original report (Cantle 2001) – white and Asian – were found to have little or no contact; therefore the term parallel lives was chosen to emphasise this. However, in Hackney, Wessendorf found an 'ethos of mixing' (2013a) and a 'civility towards diversity' (2013b) in public spaces. In the design of the book, the supplementary texts that 'stem' from the traditionally positioned field-writing are perpendicular rather than parallel. In the restaurants themselves the instances of togetherness-in-difference were less evident, but in the spaces in between and outside of the restaurants people clearly are 'rubbing along'. Thus the more traditional book setting of two columns positioned side by side in parallel wouldn't evoke this interaction, but a perpendicular setting suggests there is both a coming together and a moving away within Hackney, which evokes Wessendorf's (2013a) findings. In the majority of cases, the supplementary texts also flow across multiple pages, and therefore multiple sites within the book. So, for example, a text about kosher food cuts across the pages of a South East Asian grocers and a Vietnamese restaurant, while a text about lucky cats spans the pages of an Italian delicatessen, and a text about Shabbat dinner passes through the pages of Chinese restaurant (see Figure 5.3). This also suggests an 'ethos of mixing' and a sense that people are continually moving by and through Hackney, regardless of these particular ethnic borders that are visibly defined along the A10.

Further, more specific insights were triggered through particular eating experiences. For example, in the field-writing above, undertaken in the Turkish restaurant Şömine, I wondered if breakfast is our last remaining food-related cultural signifier. I suppose I, and many others in most large towns and cities in the UK, will often go out for a curry, a bowl of phô, or a plate of noodles in the evening or at lunchtime. However, more often than not, going out for breakfast in the UK revolves around variations of a traditional fry up, or a variety of things on toast. Similarly, in the context of our own homes, I am yet to encounter someone English who strays very far from the cereal/toast/eggs/bacon parameters with the odd croissant or pastry thrown in. A 'full English' is offered from builders' 'caffs' to gastro pubs and is variously positioned as a hangover cure, the best way to start a day's labour and a great companion to the Sunday papers. But just how English is this plate of fried eggs, bacon, mushrooms, beans (or perhaps tomatoes), black pudding (if you're so inclined) and fried bread (if you're a traditionalist and less worried about your arteries). Is it definably so? As I ate my way through Hackney

I began to see that breakfast itself might be the last remaining meal that retains some sense of cultural signification. We may be culinary explorers from midday onwards, but before then we are less likely to indulge in breakfasts that are less familiar to us, such as fit-fit, nasi lemak, or rice and banchan. Breakfast seems like a meal that remains too 'authentic' for our culturally specific tastes and is therefore one in which we remain largely indifferent and apart.

Finally, my experience of conversation and interaction in the Brazilian café led me to speculate as to whether knowledge of football (soccer) provides one with a global conversation opener. On my first visit, the TV in the café was showing a football match between France and South Africa, a match from the group stages of the 2010 World Cup in which Brazil was also participating. I asked the owner if she thought Brazil would win, and she shrugged, not seeming very confident. I suggested that according to the UK newspapers, no one in Brazil liked Dunga, the then coach of the team. 'Ah Brazilians', she replied, 'they never like the coach. One hundred and eighty million of them and they each think they know the best team'. The late Bill Shankly said that 'some people think football is a matter of life and death; I assure you it's much more than that.' Whether you believe that or not, there is no doubt that football is now a global sport. FIFA, footballs' international federation, has 211 countries as members and, from Accra to Accrington, football is avidly watched, played and discussed, night and day. The English Premier League is broadcast to every continent and almost every country and in many places all the games are live to air. With an international cast of players it is relatively easy to start a conversation – I have debated the merits of Marouane Chamakh with a Moroccan in France, discussed the career of Junichi Inamoto in Japan, marveled at the left foot of Lukas Podolski with a German traveler in New Zealand, and indulged in shared adoration of 'The Arsenal' with a plethora of locals in the Gambia. The global movement of footballers and the global reach of football as a commercial product is a clear example of the global flows and networks that interact and intersect throughout the world. In this particular footballing context, these flows facilitate further interactions and intersections as shared knowledge of the game provides an immediate link with those in, or from, far flung places. Whilst football also has its fair share of intense rivalries, many of which are violent, it would seem that it is one way of facilitating a conversation that might bring togetherness-in-difference.

Summary

Hackney is home to a diverse population, whose different communities bring with them a 'taste of home'. In this context food acts as a cultural identity marker – bringing members of the particular community together and setting them apart from others. The multitude of grocery stores, cafés and restaurants therefore offer easily identified, multi-sensory evidence of Hackney's multi-cultural population. The *Food Miles* book charts a journey through Hackney via these different premises and foodstuffs. Drawing together ethnographic field-writing, further contextual writings developed from these experiences, and documentary photography, the book utilises its four dimensions to develop a temporal and spatial reading of place. Using three different navigational systems, the book – much like place – offers multi-linear readings. The design interventions add to this, creating a space of interaction, one where the reader has to make decisions as to where their journey through the book might take them.

The book, and the process of its development has led to a range of diverse insights about Hackney in the context of food and multi-culturalism. Hackney is often discussed in terms that suggest a progressive sense of place is evident within the borough, with the diverse population coexisting happily side by side in the main. The space of the street is one in which a mix of people 'rub along' (Watson 2006; 2009) and maintain a 'togetherness-in-difference' (Ang 2003), even though the territorial borders between the Vietnamese, Turkish or Jewish establishments seem quite clearly demarcated. However, inside the restaurants the situation seems somewhat different, with many establishments seeming to evidence more of an indifference and apart scenario. This seemed to play out in those establishments that might be said to cater for food that is less familiar to Western tastes, such as the Ethiopian café, that were serving less well established local diasporic communities, and in Turkish restaurants whose cuisine and community is far more established within Hackney. Much like Wessensdorf (2013a) who found there was rarely mixing between cultures in the private space of the home, the same can be said for many of the ethnic food spaces. This, and other observations about authenticity, performativity, the cultural significance of breakfast and the potential of football knowledge to bridge cultural differences, reveal traces of the practices of everyday life and place that inform us as to how ethnicity is constructed, maintained and broached within Hackney.

Through the multi-sensory experience of food, the book constructs a space in which the reader can explore Hackney through these lenses. Yet, whilst the book utilises a range of design interventions that take advantage of its four dimensions, they remain primarily visual with some associated physical engagement. However, the book has the potential to offer a much more multi-sensory space within its pages, bringing into play both tactile and olfactory realms. The following chapter discusses a research project that explores these possibilities and develops work that goes beyond an autoethnographic perspective and engages participants in developing the narratives of place.

Bibliography

Ang, I. (2003) 'Together-in-difference: Beyond Diaspora, into Hybridity', *Asian Studies Review*, 27(2): 141–154.

Armin, A. & Thrift, N. (2002) *Cities: Reimagining the Urban*. Cambridge: Polity Press.

Cantle, T. (2008) 'Parallel Lives', in Johnson, N. (ed.) *Citizenship, Cohesion and Solidarity*. London: The Smith Institute, pp. 10–21. Available at: www.smith-institute.org.uk/wp-content/uploads/2015/10/Citizenship-Cohesion-and-Solidarity.pdf (Accessed: 10 December 2017).

Cantle, T. (2001) *Community Cohesion: A Report of the Independent Review Team*. London: Home Office. Available at: http://dera.ioe.ac.uk/14146/1/communitycohesion report.pdf (Accessed: 10 December 2017).

Carrion, U. (2001) *The New Art of Making Books*. Nicosea: Aegean Editions.

Cloke, P., Cook, I., Crang, P., Goodwin, M., Painter, J. & Philo, C. (2004) *Practising Human Geography*. London: Sage.

Cook, I. (2008) 'Geographies of food: Mixing', *Progress in Human Geography*, 6, pp. 821–833.

Cook, I., Hobson, K. Hallett, L., Guthman, J., Murphy, A., Hulme, A., Sheller, M., Crewe, L., Nally, D., Roe, E., Mather, C., Kingsbury, P., Slocum, R., Imai, S., Duruz, J., Philo, C., Buller, H., Goodman, M., Hayes-Conroy, A., Hayes-Conroy, J., Tucker, L., Blake, M., Le Heron, R., Putnam, H., Maye, D. & Henderson, H. (2010) 'Geographies of food: "Afters"', *Progress in Human Geography*, 35(1), pp. 104–120.

Crang, M. & Cook, I. (2007) *Doing Ethnographies*. London: Sage.

Del Casino, V. J. & Hanna, S. P. (2006) 'Beyond the "binaries": A methodological intervention for interrogating maps as representational practices', *ACME: AN International Journal for Critical Geographies*. 4(1), pp. 34–56.

Duruz, J. (2010) 'Floating food: Eating Asia in kitchens of the diaspora' *Emotion, Space and Society*, 3, pp. 45–49.

Garner, S. (2007) *Whiteness: An Introduction*. Abingdon: Routledge.

Gilmore, J. H. & Pine, J. B. (2007) *Authenticity: What Consumers Really Want*. Boston, Mass.: Harvard Business School Press.

Heldke, L. (2003) *Exotic Appetites: Ruminations of a Food Adventurer*. London: Routledge.

History of Hackney's Diverse Communities (2016) Available at: https://www.hackney. gov.uk/hackney-diversity (Accessed: 8 December 2017).

Hochuli, J. (1996) *Designing Books: Practice and Theory*. London: Hyphen Press.

Holtzman, J. D. (2006). 'Food and memory', *Annual Review of Anthropology*, 35, pp. 361–378.

hooks, b. (1992) 'Eating the other', in Scapp, R. & Seitz, B. (1998) (eds) *Eating Culture*. Albany: State University of New York Press, pp. 181–200.

Karaosmanoglu, D (2014) 'Authenticated spaces: Blogging sensual experiences in Turkish grill restaurants in London', *Space and Culture*, 17(3), pp. 224–238.

Kitchin, R. & Dodge, M. (2007) 'Rethinking Maps', *Progress in Human Geography*. 31(3), pp. 331–344.

Mankekar, P. (2002) '"India Shopping": Indian Grocery Stores and Transnational Configurations of Belonging', *Ethnos: Journal of Anthropology*, 67(1), pp. 75–97.

Marcu, S. (2014) 'Geography of belonging: Nostalgic attachment, transnational home and global mobility among Romanian immigrants in Spain', *Journal of Cultural Geography*, 31(3), pp. 326–345.

Massey, D. (1994) *Space, Place and Gender*. Minneapolis: University of Minnesota Press.

Mau, B. & Mermoz, G. (2004) 'Beyond looking: Towards reading', *Baseline*, 43, pp. 33–36.

May, J. (1996) 'Globalization and the politics of place: Place and identity in an inner London neighbourhood', *Transactions of the Institute of British Geographers*, 21, pp. 194–215.

Meskimmon, M. (2003) *Women Making Art*. London: Routledge.

Petonnet, C. (1982) 'L'observation flottante: L'exemple d'un cimetière Parisien', *L'Homme*. 22(4), pp. 37–47.

Pink, S. (2015) *Doing Sensory Ethnography*. 2nd edn. London: Sage.

Policy and Partnerships Team (2017) *Facts and Figures Leaflet*. Available at: www.hackney. gov.uk/population? (Accessed: 8 December 2017).

Richardson, L. & Adams St. Pierre, E. (2005) 'Writing: A method of inquiry', in Denzin, N.K. & Lincoln, Y. S. (eds) *Handbook of Qualitative Research*. Thousand Oaks: Sage, pp. 923–948.

Shields, R. (1991) *Places on the Margin: Alternative Geographies of Modernity*. London: Routledge.

Sinclair, I. (2009) *Hackney, That Rose Red Empire: A Confidential Report*. London: Penguin.

Sutton, D. E. (2010) 'Food and the Senses', *Annual Review of Anthropology*. 39, pp. 209–223.

Timur Öğüt, Ş (2008) 'Turkish restaurants in London: An ethnographic study on representation of cultural identity through design', *A|Z ITU Journal of the Faculty of Architecture*, 5(2), pp. 62–81. Available at: www.az.itu.edu.tr/az5no2web/07timur 0502.pdf (Accessed: 10 December 2017).

Vertovec, S (2007) 'Super-diversity and its implications', *Ethnic and Racial Studies*, 30(6), pp. 1024–1054.

Watson, S. (2009) 'The Magic of the Marketplace: Sociality in a Neglected Public Space, *Urban Studies*. 46(8), pp. 1577–1591.

Watson, S. (2006) *City Publics: The (dis)enchantments of Urban Encounters*. Abingdon: Routledge.

Wessendorf, S. (2013a) 'Commonplace diversity and the "ethos of mixing": perceptions of difference in a London neighbourhood', *Global Studies in Culture and Power*. 20(4), pp. 407–422.

Wessendorf, S. (2013b) *Commonplace diversity: stable and peaceful relations across myriad differences in Hackney*. Available at: http://blogs.lse.ac.uk/politicsandpolicy/ commonplace-diversity-stable-and-peaceful-relations-across-myriad-differences-in-hackney/ (Accessed: 8 December 2017).

Wolfinger, N. H. (2002) 'On writing fieldnotes: Collection strategies and background expectancies', *Qualitative Research*. 2(1), pp. 85–95.

Zawieja, J. (2009) 'Imagine a house: Transcript', in Naik, D. & Oldfield, T. (eds) *Critical Cities: Ideas, Knowledge and Agitation from Emerging Urbanists: Volume 1*. London: Myrdle Court Press, pp. 138–145.

6 Home-making, memories and materiality

Stuff

Introduction

As we discussed in Chapter 1, notions of home were of central concern to humanistic geographers, and whilst their romanticised version of home as a private haven has since been critiqued by feminist geographers amongst others, home still remains an important site of study for academics within cultural geography and anthropology, particularly in relation to material culture. Tuan suggests 'we are what we have' (1980: 472) and there is no doubt that for the majority of us, the houses, apartments or rooms we live in are filled with 'stuff'. Contrary perhaps to the understanding of the external viewer, this is not just any old stuff; it is *our* stuff. Whilst we undoubtedly accumulate items that enable our lives to function more comfortably, sustainably, or seamlessly, the stuff that is really important to us is that to which we have an emotional connection. It is predominantly made up of 'keepsake' objects – the kind of things that we literally keep over long periods of time. The phrase 'an accumulation of stuff' seems to imply there is no-one involved, that the stuff increases imperceptibly. Ben Highmore suggests that 'things turn towards us: they call us, sidle up to us' and so we 'attach ourselves to the thingly world' and 'surround ourselves with keepsakes and mementos' (Highmore 2011: 58). Such objects are evocative; they are 'companions to our emotional lives' (Turkle 2011: 5) and can hold unexplored worlds, containing within them 'memory, emotion, and untapped creativity' (Pollack 2011: 228).

> My grandmother has been dead for nearly fifteen years, but when I make cookie dough with my children I use her wooden rolling pin with its chipped red handles . . . This tactile ritual takes me back to the warmth of her kitchen, the aromas of her cooking, and the comfort of her presence . . . As I use her rolling pin and feel its texture and weight against my floured hands, I think of the hundreds of pies and cookies it helped create. It anchors me in the past, yet continues to create memories for the future. The object becomes timeless.
> (Pollack 2011: 227)

Whilst this memory is articulated through words, the trigger is a material object. Our relationship with things is one that engages all the senses, not just the visual; we pick them up and handle them, feel familiar worn grooves or textures;

we smell them, drinking in reminiscences of particular moments; and we listen to them, placing a song at a moment in time. Chapter 4 discussed geographers' explorations of the evocative potential of creative writing (Lorimer 2008: 182), and there is no doubt that such writing, as the quote above aptly demonstrates, can move us in ways that resonate with our own pasts. However, this type of representation reduces the materiality of these objects, and our embodied, multi-sensory encounter with them, to words. This chapter focuses on the creation of a book – *Stuff* – that provides the reader with a space in which they are able to engage with material objects and handle items that, alongside the written content, will transport them to their own houses, their own things and thus, in turn, their own memories.

Unlike the previous chapter in which the research was articulated predominantly through auto-ethnographic means, this project is participatory in nature, drawing together participants' memories of 'stuff' and a life story contextualised through one participant's practice of collecting. Whilst the development of this project draws from the framing of a book as a temporal space as discussed in the previous chapter, it also draws from Arnar's (2011) discussion of the 'livre d'avant garde' – an experimental book that is distinct from the artists' book or 'livre de peintre'. A less well known term these days than artists' book, the livre d'avant garde was seen as more radical than the livre de peinture, in that its ambitions were to challenge the conventional form of the book with a view to challenging both art and life (Arnar 2011: 2). Stuff therefore challenges Ingold's (2007: 24) assertion that the page has been silenced by the advent of mechanical print and engages with Pink's concerns as to the challenges of representing embodied, multi-sensory experiences (Pink 2015: 40). In doing so, it positions the book as capable of creating an experience that engages multiple senses.

Home

Much like place, home is a complex theoretical concept, and like place, one's sense of home will vary depending on one's perspective and experience. In some contexts, home simply means a house, bricks and mortar, or some kind of physical structure. In others, home centres on relationships and connections that span both space and time. Ideas of home are often associated with positive emotions and warm, nostalgic memories of childhood. However, home can also inevitably produce negative feelings, or a mixture of the two, and how one defines and makes home is likely to be, informed by both our past and present experiences and how we wish our future to unfold (Blunt & Dowling 2006: 1). So, whilst a home is a physical shelter it is also much more than this, and the idea of home is one that is filled with emotional meaning (Csikszentmihayli & Rochberg-Halton 1981: 121). A sense of belonging or attachment is also linked to the idea of home, yet in a world when lives are often more transient than in the past and where transnational migration is commonplace, 'home' also stretches beyond the boundaries of a single dwelling in a single location. Home is therefore simultaneously both 'material and imaginative' (Blunt & Dowling 2006: 22) and can be described as: 'a spatial imaginary: a set of intersecting and variable ideas and feelings, which are related to context, and which construct places, extend across spaces and scales, and connect places' (Blunt & Dowling 2006: 2). Home could therefore be described as being a place or site; a set of feelings, emotions and meanings; and the relations between this material and affective space (Blunt & Dowling 2006: 22).

As we have discussed previously in Chapter 1, home was a central focus for humanistic geographers in the 1970s and 1980s, and at that time was 'cast as a uniform space of safety and familiarity' and 'a site of authenticity and experience' (Brickell 2012: 225). Humanistic geographers were particularly interested in the way people develop a sense of home and, through this, a sense of belonging. Home was seen as a meaningful place, one that formed 'an irreplaceable centre of significance' (Relph 1976: 39) and in a world that was becoming increasingly globalised and 'placeless' (Relph 1976), home was therefore seen as affording both a sense of place and belonging. This conception of home was one built on stasis and permanence that offered a respite from the ever-changing world that threatened people's ability to create home as a form of haven or retreat from modern life (Blunt & Dowling 2006: 14). This view of home offers a perspective that has since been revealed as romanticised and idealised, and as leading to 'a normative association between home and positivity' with home becoming a 'metaphor for experiences of joy and protection' (Brickell 2012: 225). For example, ideas of home as a 'haven' still largely neglect women's experience of home. For men, home might be a sanctuary that offers a retreat from society and a refuge from the daily grind of the workplace, but for women, the home is often the site of domestic labour and a never ending daily set of household chores (Blunt & Dowling 2006: 16). Similarly, as we discussed briefly in Chapter 1, a woman's experience of the home can be oppressive and it may be the site in which she experiences domestic violence (Rose 1993: 12). Therefore, gender is a key lens through which to critically consider home, the experience of home and home-making practices. This failure to engage with gender is one example in which humanistic geographers' notions of home display a lack of understanding of the co-constitutive role played in the construction of 'home' by social structures within society and experiences of place. Much like the contemporary understanding of place which now accepts it as being both local and global, home is now recognised 'as a process of establishing connections with others and creating a sense of order and belonging as *part of* rather than *separate from* society' (Blunt & Dowling 2006: 14; italics in original).

Homes, therefore, do not simply exist, they are constructed via everyday practices (Dowling & Power 2013: 294) – they are 'lived' and are therefore continually re-constructed (Blunt & Dowling 2006: 23). Such home-making practices incorporate both the imaginative and the material – we thus create home through the formation of social and emotional relationships (Blunt & Dowling 2006: 23) and through our use, positioning, and relationships with objects, furnishing, and décor (Dowling & Power 2013: 294). The study of the material environment of the home and a return to an interest in this aspect of home-making practices has both contributed to, and developed from a 'material turn' within geography (Blunt &

Dowling 2006: 74). However, the objects that we choose to have in our homes do not just resonate in terms of their aesthetics, they also offer both temporal and spatial connections to people, places and moments from our past lives.

Home-making, memory and materiality

We fill our houses with many different types of materials – from fixtures, fittings and electrical devices to textiles, décor and ornaments. In simplistic terms, we choose to furnish and decorate our houses with things we like in order to 'create a sense of "homeyness" and belonging' (Dowling & Power 2013: 294). Seen individually, or even en masse, by an outsider, these materials are quite likely to seem unremarkable, but together they transform 'our house into our *home*' and can be seen as 'a material testament of who we are, where we have been, and perhaps, even where we are heading' (Hecht 2001: 123: italics in original). The 'things' that make up our stuff are actually much more meaningful than the word implies. 'Thing' suggests something we can't be bothered to name properly, or something we don't know, or can't remember, the name of. However, in the context of our homes, *our* things are resonant, they are 'invested with meaning and memory' and trace the journey of our lives, whilst both 'framing and reflecting our sense of self' (Hecht 2001: 123). We know them intimately; rather than being a 'thing' we have forgotten the name of they are a thing that helps us remember; they are things that can help us travel through both space and time.

> It is here, still closed in front of me. This is my grandmother's suitcase, one she would have used when she came from France. It is small, just large enough for one person to pack for a one- or two-week trip. It is firehouse red . . . Two and a half years after I packed the suitcase I begin to open its buckles, one at a time . . . I am not ready for the smell of her perfume, her hair, her jewelry, and clothes to come at me so fast. She reaches me from inside . . . The suitcase brings her back to me with the worry that I will lose her if I open the suitcase too often; her smell will evaporate, the letters will fade, and the clothes will no longer hold her shape.
>
> (Dasté 2011: 246–249)

Thus, home becomes our 'private cosmos' (Hecht 2001: 123) in which the collection of material effects creates an environment that embodies what is significant to us (Csikszentmihalyi & Rochberg-Halton 1981: 123). In this context, home becomes a shelter for the stuff that has shaped our lives and that makes our lives meaningful (Csikszentmihalyi & Rochberg-Halton 1981: 139). Even seemingly insignificant items such as torn bus tickets can trigger evocative memories that 'promote emotions out of all proportion to their material paltriness' (Cardinal 2001: 23). So, 'homeyness' isn't simply about comfort and warmth in a physical sense, it is as much – if not more – about a psychological sense of home. Thus, whilst a chair may offer a visitor a comfortable seat, for its owner it may also offer a window to a particular time, place or person. Whilst the psychological importance of such items is therefore lost on those who have no connection to them, we also often use, arrange and position these items in ways that highlight their significance to us (Dowling & Power 2013: 294).

Pearce (1994: 27) suggests that the past survives in three different ways: through objects and material culture; through the physical landscape; and through narrative. One of the key ways we use such items is as aide memoires for storytelling, and in this context, the stories we tell of ourselves. The human species has been described as homo narrans (Niles 1999) and what defines homo narrans is the ability to imagine and inhabit spatial and temporal worlds beyond the present reality (Niles 1999: 3). Our desire to tell stories seems inherent; imagining life without recounting tales of our day to friends or reminiscing about past events is almost impossible, and it is through such storytelling that we 'possess a past' and that we articulate our values (Niles 1999: 2). Storytelling is therefore one way in which we construct, develop and maintain a sense of identity that is 'woven out of memories and experience' (Hecht 2001: 129). However, although storytelling is key, this is not to suggest the process of preserving and presenting our life histories and memories is purely linguistic. Our experience of the world is multi-sensory, and as we discussed in Chapter 4, our senses don't operate individually, rather they combine to give us a coherent multi-sensory experience of the world. Thus the physical presence of objects from our past that we can listen to, touch, smell, and view acts as 'sensory evocations' of our past memories and helps us feel at home in the present (Hecht 2001: 141). Such personal objects 'speak of events that are not repeatable, but are reportable'; they authenticate the narrative we construct around them (Pearce 1994: 196). In this sense, these objects function as 'prostheses' of the mind (González 1995: 133). Souvenirs and mementos in particular are key here; the things we keep specifically because they move us in some way or are significant to us for a particular reason. In reminding us of a holiday long ago, a long-deceased grandparent, or childhood summers, they 'have the power to carry the past into the present' and 'are samples of events which can be remembered but not re-lived' (Pearce 1994: 195). They 'are lost youth, lost friends, lost past happiness; they are the tears of things' (Pearce 1994: 196). We know these things intimately, they are part of the fabric of our being and are autobiographical in that sense. Such a 'material memory landscape' forms a 'spatial representation of identity' – an 'autotopography' (González 1995: 133). An autotopography does not include all our possessions, just those that 'signify "individual" identity'. However, it does not have to be a carefully chosen and arranged display of mementos on a shelf or mantelpiece, it may be a jumble of objects contained in a box in a dusty attic (González 1995: 134). Whichever form it takes, the autotopography creates a 'physical map of memory, history and belief' and becomes 'an addition, a trace, and a replacement for the intangible aspects of desire, identification, and social relations' (González 1995: 134). This representation of the self draws together objects that offer a multi-sensory topography that provides a metonymic strategy through which we are able to view our past life (González 1995: 134).

The home, and rooms within it, can therefore be seen as similar to a museum in that it holds displays of material artefacts that we have carefully curated in order to construct and reproduce memories. However, unlike a museum which assists in the construction and narration of national identity in a public context, the home constructs, reproduces and narrates the 'personal memories and familial identities' of its occupants (Meah & Jackson 2016: 527). The Romans developed a process for remembering – the method of loci – based on visualisation of spatial memories within the layout of a building, and Eco, when discussing memory suggests that: 'Remembering is like constructing and then traveling again through a space. We are already talking about architecture. Memories are built as a city is built' (Eco 1986: 89). An autotopography therefore functions similarly, with the individual objects acting as mnemonic devices that '*create a space* for the memory they represent' and for a moment we are able to reconstruct the past in the present to 'convince ourselves of something we wish to be or have been' (González 1995: 135–136; italics in original). Therefore, memory is not just reflective, but also constitutive (González 1995: 136) and objects, then, can be described as having agency – the object is an 'external force' through which memory is 'activated' (González 1995: 135).

Memories are also triggered both involuntarily and intellectually and in the context of autobiographical objects, what is most often activated is an involuntary memory. The term – mémoire involontaire – was originally coined by Proust (2013) and captures the experience of suddenly remembering something vividly that one had not deliberately intended to recall. For Proust, this type of involuntary memory contains an 'essence' of the past and is best described by his memories triggered by a madeleine dipped in tea. Proust's reminiscences and others like them, including our own, might be said to be nostalgic. Autobiographical objects such as souvenirs have been described as having their 'roots in nostalgic longing for a past which is seen as better and fuller than the difficult present' and this creates a backwards spiral in which the original experience that is being evoked happened longer and longer ago and so connecting with it can only be achieved by 'building up a myth of contact and presence' (Pearce 1994: 195–196). The word nostalgia comes from the Greek roots 'nostas', which means return home, and 'algia', which means longing. Although it has its roots in ancient Greece, the word nostalgia wasn't actually coined until the seventeenth century and was used in a medical context, relating to home sickness. Today, it can be defined as a longing for a home that no longer or exists (or perhaps never existed) and although we often therefore align it spatially, it is perhaps more accurately a longing for a different time – often that of our childhood or formative years (Boym 2007: 8). Boym (2007) distinguishes between two main types of nostalgia – restorative and reflective – with restorative emphasising the nostos or home, and reflective thriving on the algia or longing. It is reflective nostalgia that is more at play in autobiographical objects, as it engages with an individual narrative and 'does not follow a single plot but explores ways of inhabiting many places at once and imagining different time zones' (Boym 2007: 13). In this context, the reflective nostalgia triggered by our souvenirs and family heirlooms is both an absence and a presence; the object is there, along with the memory, but the past is no longer a place we can return to. Nostalgia by its very nature can never be satisfied, as 'it is the longing that structures this desire' – 'nostalgia is always a desire for desire itself' and the 'gap between signifier and signified, between the construction of a narrative and its referent, can never really be crossed' (González 1995: 137).

Nostalgia is often criticised for its part in the re-creation of an idealised past that more often than not reflects a dominant narrative that has been expunged of any contentious and contested elements and is therefore presented as overtly sanitised. For example, nostalgic views are often those associated with regressive views of place, and whilst this shouldn't be downplayed, we also need to recognise that memories are inevitably selective in any context and at any scale. In respect of reflective nostalgia, a grand historical narrative applicable to a nation isn't what is at stake. Rather, 'The focus here is not on the recovery of what is perceived to be an absolute truth, but on the meditation on history and the passage of time' (Boym 2007: 15). For those who are unable to physically travel, this type of memory performs an important role. For example, Rose (2003) found that collections of photographs enabled women to 'stretch' both space and time beyond that of the present home and in doing so connect with family and friends in other places. This kind of travelling would work for those of all ages, but for older people particularly, homes are multi-scalar and autobiographical objects and spaces 'have their own agency' and can be '*constitutive* of the social time and space of later life' rather than simply being 'passive props' for reminiscence work (Hockey, Penhale & Sibley 2005: 135; italics in original). Such memories also involve a form of active physicality, even for those who might be predominantly housebound, as they are 'sourced through our bodies interactions between haptic perception, the senses, tactile experiences, and movement' (Bhatti et al. 2009: 71–72). So, whilst one's limited mobility might mean one's world might be reduced to one's home, and perhaps even one room or one's favourite chair within that, our 'stuff' still connects us to people and places beyond this.

For the person who owns the meaningful object, this agency or power is unique, however, for the stranger the same object is likely to be meaningless and inert. The past that is conjured up by souvenirs and mementos is an intensely individual one – 'no one is interested in other people's souvenirs' (Pearce 1994: 196). This is perhaps most evident when viewing possessions resulting from a house clearance in the context of a junk shop or auction, or even when clearing the homes of deceased relatives. Even for family, whilst some things will be familiar and evocative, others will be devoid of meaning: 'Even the most admirable and best-spent lifetime seems to peter out in a residue of old coats and shoes, stacks of faded newspapers, and all kinds of memorabilia lodged in jars, boxes, chests and cupboards, their significance poignantly annulled' (Cardinal 2001: 23). Borges' (1977: 57) poem *Inventory* captures the experience of entering the unfamiliar storage space of an attic and encountering a jumble of seemingly disconnected, meaningless fragments of someone's life. The inventory records an eclectic list of items including 'A Paraguayan hammock with tassels, all frayed away. Equipment and papers'; 'A clock stopped in time, with a broken pendulum'; 'A cardboard chessboard, and some broken chessmen'; and, 'A photograph which might be of anybody'. Borges' describes these objects as 'the flotsam of disorder' – to anyone but the owner of these objects little makes sense and questions remain unanswered – did they visit Paraguay and bring home a souvenir of their trip; is the photograph a relative or friend; might I be distantly related to this person? It is unlikely we will ever know the answers as a 'whole context of richly layered experience has simply dropped away without recall' (Cardinal 2001: 24). We all have to acknowledge that our possessions 'are doomed to be classed one day under the ultimate heading: EPHEMERA' (Cardinal 2001: 30; emphasis in original).

Yet some possessions escape this seemingly inevitable demise, and do offer the outsider some visible sense of order, if not the associated memories and emotional connection. Collections can be seen as self-contained entities within our 'stuff'. Unlike the accumulation of mementos that encapsulate meaningful moments but can take any form, a collection has clear parameters defined not only by the collector, but by the elements of the collection itself. Collecting might be described as a type of consuming. However, unlike the literal meaning of consumption which is to use up or devour, collecting 'is about keeping, preserving, and accumulating' (Belk 2006: 534). Belk defines collecting as: 'The process of actively, selectively, and passionately acquiring and possessing things removed from ordinary use and perceived as part of a set of non-identical objects or experiences' (Belk 1995: 67). Unlike the imperceptible accumulation of 'stuff', therefore, collecting and collections are purposeful acts and collecting can be seen as a way of obtaining a sense of control in our increasingly complex lives (Belk 1988: 154). The collection, unlike the souvenir or memento 'expresses classification in place of history' and 'replaces temporality with order' (Shelton 2001: 13). Collections are often put on display within the home, with mantelpieces, windowsills and shelving acting like display cases in a museum within areas of the house that are 'open to the public'. It is because of this order and definition, that collections retain some meaning for those who don't own them – we can see the parameters and understand the logic – we can rationalise them. Yet many collections begin because a specific item has been bought or received as a gift that has some emotional or personal significance. Therefore, collections often extend beyond the rational and may have more links to history, temporality, emotions and memory than we might think.

A geo/graphic approach to home, materiality and memory

Whilst the initial section of this chapter has outlined the theoretical context for the project, as with the previous two chapters, the development of *Stuff* was not in response to this theoretical investigation; it was triggered by ongoing research into life in the London borough of Hackney. As part of this process I developed a cultural probe pack which was distributed to 33 residents within the borough.

Cultural Probes

In discussing the recording of everyday life and place, Highmore (2002: 171) suggests a tool-kit is needed that enables the 'different registers of a polyphonic everyday to be heard' and this could perhaps be likened to a description of cultural probes. Originally developed as strategy to assist researchers pursuing responsive approaches to experimental interactive design (Gaver, Dunne & Pacenti 1999) the first probe packs were designed to include a set of postcards, maps, a disposable camera, a photo album and media diary. Cultural probes have since been used in a number of disciplines, including ethnographic research (for example, Robertson 2008), and the probes are designed to be fun to use, to break down the barrier between researcher and participant, and to provide inspirational responses rather than information. They are primarily designed to understand 'people in situ, uniquely, not abstractly en masse' (Hemmings et al. 2002: 50). The specific contents of the pack are not defined, but can be designated for each individual research project and therefore, are able to relate to the specifics of the study and participants involved. The method has been developed from the perspective and traditions of the artist/designer, rather than the scientist, and has its roots in the approach and ideas of the Surrealists and Situationists (Gaver, Dunne & Pacenti 1999, Gaver et al. 2004). The designers themselves state that: 'Scientific theories may be one source of inspiration for us, but so are more informal analyses, chance observations, the popular press, and other such 'unscientific' sources' (Gaver, Dunne & Pacenti 1999: 24). The probe returns are used in an 'openly subjective' way, stimulating the designers' imagination, rather than defining a particular issue that 'needs addressing' (Gaver, Dunne & Pacenti 1999: 25). Many of the disciplines that have adopted cultural probes as a method have taken this inspirational and openly subjective approach to be a failing, and have attempted to rationalise the probes, asking specific questions and analysing results (Gaver et al. 2004). This misses the point of the probes – their values are 'uncertainty, play, exploration, and subjective interpretation' (Gaver et al. 2004: 53) – they, and the researchers who developed them, embrace subjectivity rather than seek to minimise it. However, it is unlikely the returns were, as Gaver et al. state (2004: 53) not analysed at all; rather they will have been analysed as part of the design process. In order to generate ideas, by necessity the designer has to engage with the research materials cerebrally. As celebrated designer Paul Rand states: 'In order, therefore, to achieve an effective solution to his problem, the designer must necessarily go through some sort of mental process. Consciously or not, he analyses, interprets, formulates' (Rand 1970: 12). If there were no evaluation or analysis of the materials generated by the probes, it is unlikely designers could use them for inspiration and generate ideas from them. Gaver et al. (2004: 56) state that the relationship between the probes and the design process is complex, whilst Boehner et al. (2007: 5) suggest that much of the literature about the broader use of the probes shows a lack of detail when discussing how designers move from probe to design.

Perhaps what is partly at the root of this issue of defining the probes as resistant to analysis is the perception of what 'analysis' is. I would suggest that in order to differentiate their method from more traditional social science tools such as questionnaires, Gaver et al. (1999, 2004) deliberately sought to position the probes firmly within an art and design context. One way of reinforcing this position was perhaps to deliberately eschew the process of analysis in relation to the probes, thus further encouraging an 'art not science' binary opposition. However, this position is perhaps established in something of an artificial way, as Gaver et al. (2004) seem to assume a very narrow, positivist definition of research and analysis, one that could 'be conceived primarily in terms of data handling' (Coffey & Atkinson 1996: 6). As we have seen in Chapter 3, contemporary social scientists are now engaging in a wide range of creative methods that make such a distinction almost redundant, and see analysis as 'essentially imaginative and speculative' (Coffey & Atkinson 1996: 6).

Probe returns also encourage the construction of narratives about the participants, and as with many of the tactics and approaches discussed previously, they make the familiar strange and the strange familiar, but they don't dictate what should be developed through the design process (Gaver et al. 2004). With their philosophical approach rooted in the disciplines of art and design, and their ability to provide 'fragmented illustrations and narratives' (Jääsko & Mattelmäki 2003: 4–5) the probes are an appropriate method with which to explore everyday life and place. The fragmented nature of the narrative has parallels with Massey's definition of place, and offers the designer an opportunity to develop a fuller narrative from the responses. As Gaver et al. (2004: 55) state: 'Rather than producing long lists of facts written about our volunteers, the probes encourage us to tell stories about them, much as we tell stories about the people we know in daily life'.

In this research, participants from Hackney were recruited through a variety of ways; many responded to calls sent via emails to organisations within the borough, such as Hackney Silver Surfers, various tenants' and residents' associations, and the Hackney Society. Other participants received the email as it was forwarded to them by someone in one of these organisations. A further five participants were recruited through my own personal contacts within Hackney. Of the 33 packs sent out, 27 were completed and returned. The probe packs in this instance consisted of a series of postcards with questions on the reverse, a 12 shot disposable camera, a memory/story sheet, a recipe sheet and a journey log. The questions on the postcards were as follows: 'What does Hackney sound like?'; 'What is your favourite journey in Hackney?'; 'What makes your house a home?'; 'What is the best part of your day in Hackney?'; 'What don't you like about Hackney?'; 'How has Hackney changed?'; and 'What is the best thing about Hackney?'. In relation to the question 'What makes your house a home?', one answer in particular caught my eye.

> All my 'Stuff' I suppose, of which there is a great deal. Never one to do anything by halves and having taken up many projects (most unfinished), Stuff has proliferated to a great extent, all considered as 'stocks' that could and would be used at some time or other. These include beads, fabrics, wool and artists' materials. What will my children do with all of it when I peg out.
> (CL 2009)

Inspired by this particular response, and two other similar ones, I contacted these individual participants to see if they would be willing to talk about this 'stuff' further with me.

Adapting research methods

Gaver, Dunne & Pacenti's (1999) original cultural probe studies did not include post-return interviews and researchers who have developed more sustained contact with participants via interviews have been criticised for destroying the idea of the probes as functioning as inspiration rather than information (Gaver et al. 2004). However, Robertson (2008: 4–5) argues that, although contentious, it can be profitable to adapt the probes and seek a balance between inspiration and information by utilising more traditional methods such as interviews alongside the probes. Boehner et al. (2007: 8) also state that 'there is nothing wrong with adapting probes', but what is problematic is adaptation without reflection – the researcher needs to think through why and how the new variants make sense, with an awareness of implications for the research. Patton (2002: 52) suggests that interviews further enable the development of a non-judgemental, empathetic understanding, thus giving an empirical basis for research from which it is possible to portray the perspectives of others. In this context, I felt that not to discuss these three specific probe returns further would actually restrict the potential of the project.

Each of the three participants were willing to discuss their responses further and not only was CL willing, but a few days later I received an unsolicited five-page Word document in my inbox entitled 'The stuff of dreams' – essentially a life story charted through CL's various collections and craft interests. We then continued our contact with a wide ranging, unstructured conversation about this 'stuff' that took place at CL's home. This conversation was prompted by the life story and by the sight of much of the 'stuff' on shelves and dressers. I carried out similar conversations with two other participants who also shared memories and stories of particularly significant items. The three participants also allowed me to photograph many of the items they had referred to specifically in our conversations. This extended and personal contact enabled a particularly appropriate perspective to be developed and re/presented through the project, and although it could be said to contravene the original intention of the probes, the research was enriched by this and the responses provided further inspiration for the continued development of the project.

Re/presenting place

The probe responses and subsequent conversations led me to write a more 'academic' response to the subject in the form of an essay to contrast with, and further contextualise, the personal narratives. Thus I had four different types of material – the life story; the conversations and memories; the essay; and the photographs. As some of these texts were relatively extensive, once again a book seemed the most obvious vehicle in which to explore the re/presentation of this particular experience of home, and as previously, the design of the book sets out to challenge Ingold's (2007) perception of the page as silenced. Stemming from an understanding of the book as unfolding in a temporal dimension as discussed in the previous chapter (Mau & Mermoz 2004: 33; Carrion 2001; Hochuli 1996: 35) this project engages with both form and content to produce what Drucker calls the 'phenomenal' book, 'the complex production of meaning and effect that arises from a dynamic interaction with the literal work' (2003: no pagination). She describes the book as 'a dynamic interface, a structured set of codes for using and accessing information and navigating the experience of a work. Books are immersive, absorptive, complex' (Drucker 2004: vii). A book is effectively co-created through the reader's interaction, 'it is not an inert thing that exists in advance of interaction, rather it is produced anew by the activity of each reading' (Drucker 2003: no pagination) – much like the contemporary view of a map discussed in Chapter 2 (Del Casino & Hanna 2006; Kitchin & Dodge 2007). However, this view of the reader's role in the 'production' of the book is not simply the logical extension of poststructuralist theorising, it is a view that has been held much longer than the past few decades.

Stéphane Mallarmé, a nineteenth-century poet, writer, and designer, believed that a book could redefine reading, and the role of readers, through the exploitation of its 'textual, visual and temporal elements' (Arnar 2011: 2). Indeed, Barthes 'essentially credits Mallarmé with having given birth to the reader', because through his work readers became active participants in a creative reading process (Arnar 2011: 287). Mallarmé developed this approach at a time when the publishing industry was suffering a financial crash due to competing newer forms of media such as photography and newspapers (Arnar 2011: 47). However, he used this challenge to re-evaluate books, understanding that, rather than lead to their extinction, the new forms could prompt an evolution of books (Arnar 2011: 291). There are perhaps parallels here with the current move within the social sciences, and geography in particular, towards creative methods using newer technologies such as sound or film and away from textual and print based forms of representation.

As discussed in the previous chapter, many of the design strategies chosen were adopted in order to avoid a 'confused presentation' (Crang & Cook 2007: 201); to utilise the four dimensions of a book and maximise the potential in its material form; and to work with both form and content in such a way as to offer the reader an engaging experience that is sometimes lacking in ethnographic work (Richardson & Adams St. Pierre 2005). So, each of the different types of texts is set in a different typeface to enable the reader to distinguish between them and thus the different voices and contexts; the book once again uses French folds in which to position 'hidden' elements; and typographical interventions throughout the book also engage the reader with the content in both conceptual and physical ways. For example, when the essay about 'stuff' refers to a collector nearing completion and experiencing a fear that prompts her to redefine the task, the text begins to adopt a different system of positioning. The blocks of text are turned by 90°, and by the end of the paragraph, sit outside of the columns of the grid, implying the shift in focus of the collection (see Figure 6.1). In the final section of the essay, that centres on possessions being lost upon death, the type starts to break out of the grid and away from the horizontal baseline (see Figure 6.2), thus implying a pile of disordered clutter. Throughout the book, the participant's life story is typeset at a 90° angle, so readers must turn the book to read it. Like the previous chapter, this physical act once again suggests that the readers literally move away from the other text, reorienting themselves through this new information – perhaps like turning a map so it points in the direction one is going (see Figure 6.3). The participant's life story includes several self-deprecatory asides, not central to the story, but key to gaining a sense of the participant's tone of voice and sense of humour. These asides are used in a way that visually references a scribe's 'gloss' to a manuscript text. A series of glyphs indicate the point in the text that relates to the particular aside, but unlike the present footnote system that gathers all the relevant notes together at the foot of a page or the end of a chapter, these asides are positioned throughout each page. This requires the reader to 'travel' the page in order to link the two symbols together (see Figure 6.4). The texts that explain the particular significance of many of the images of items contained in the book are hidden within French folds, behind the image they refer to. This engages with the idea that one's precious possessions are often meaningless to others. By positioning the captions in this way, the reader sees a seemingly ordinary, old chair at first, with no sense of why it is meaningful and what significant memories might be associated with it. By going beyond the face value of the image, by literally

Figure 6.1 Blocks of text break out of the grid implying a shift in the focus of a collection

All of our lives will inevitably 'peter out in a residue of old coats and shoes, stacks of faded newspapers, and all kinds of memorabilia lodged in jars, boxes, chests and cupboards, their significance

Cardinal 2001:23   poignantly annulled'. The stuff that once held significance for us and created our 'physical map of memory, history and belief' reverts to be just stuff.

Its inability to communicate beyond the individual renders it mute, it becomes confined to its material reality, still three-dimensional, but in this respect a one-dimensional signifier, incapable of enlightening us of the memories and meanings that once were ascribed to it.

Figure 6.2 Lines of text break out of the grid implying a pile of disordered clutter

looking behind the surface of the page and discovering the captions, the chair becomes a gateway to memories of moving to, and falling in love with, a new city (see Figures 6.5 and 6.6).

Whilst these design interventions undoubtedly contribute to 'the complex production of meaning and effect' (Drucker 2003: no pagination) through the reader's interaction with the book, other interventions explore an embodied sense of home and place, engaging with multiple senses in a deliberate attempt to move

Figure 6.3 Participant's life story set at a 90° angle

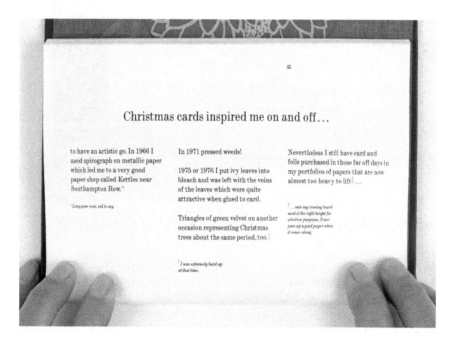

Figure 6.4 Glyphs indicate the points in the text that relate to the particular asides

Figure 6.5 A participant's chair that signified moving to, and falling in love with, London

Figure 6.6 Caption within French fold to reveal the meaning behind the chair

from an affective experience to an emotional one. For example, the book has been deliberately designed to be small (14 x 18cm) in format. Therefore the reader is able to develop a close physical relationship with it as the volume is easily held within their hands. The book's size reflects a close-up, intimate view of place. It does not, however, immediately reveal its contents or that it is anything other than traditional in nature. The cover and bindings are straightforward, with the title embossed on buckram cloth (see Figure 6.7). The cover acts as a front door, a threshold between public and private space, and on 'entering' the book, readers encounter end pages made with brightly coloured wallpaper (see Figure 6.8). Further non-traditional materials are used to maximise the engagement of multiple senses − in particular touch and smell − and trigger specific reminiscences. Glassine paper interleaves some of the pages that contain images (see Figure 6.9), and perfumed drawer liners create other pages (see Figure 6.10). The use of

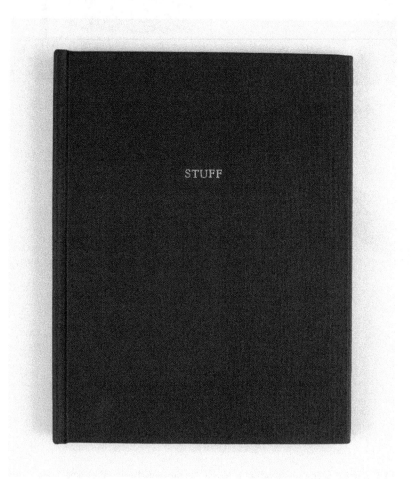

Figure 6.7 Traditionally embossed front cover

Figure 6.8 End pages made from wallpaper

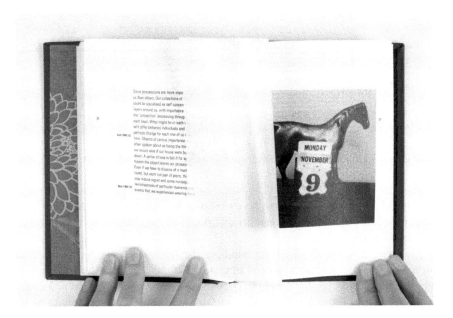

Figure 6.9 Glassine paper covers many of the key images

these materials draws the reader's imagination to sites and objects they may have experienced previously: homes of older relatives or old family photographs in a

Figure 6.10 Perfumed drawer liners are used to create some pages

traditional album collected through generations. This prompts the reader to frame their own understandings and memories of such items within the context and content of the book.

In developing these ideas and handling the materials, my thoughts drifted to my own family photographs and ornaments. In this sense, as in the previous chapter, I was reimagining place (Pink 2015: 143) through this design and material led inquiry. I thus began to think of *Stuff* in terms of a collection of pages bound together – individual parts that create a whole – much like the idea of an autotopgraphy (González 1995). This opened up the potential for the inclusion of other materials and ephemeral elements of a non-traditional nature, offering readers additional tangible references to such collections. Bound within *Stuff* are items that function as another page of the book and create separate spaces of exploration, discovery and imagination. Envelopes containing used stamps, cigarette cards, letters, old photographs and slides – purchased from second hand shops in Hackney – are included, allowing readers to draw out the contents (see Figures 6.11 and 6.12). This physical engagement triggers readers' memories of

Figure 6.11 Envelope containing cigarette cards and stamps

Figure 6.12 Envelope containing old cards and letters

their own childhood hobbies, family holidays, old friends, or the act of looking through drawers and cupboards in family homes that contain such collections. The book becomes interactive and, with readers at the heart of the process, enables them to bring thoughts, memories, and emotions of their own to the experience that reinforce the possibility that each reading of the book will become an individual, personal journey. By introducing elements that engage the reader through multiple senses, the book, much like our engagement with such objects themselves, becomes an embodied experience. The material form of the book thus places an equal emphasis on the embodied dimension as it does with the traditional textual and linguistic dimension. However, as with the previous chapter the form of the book and its re/presentational qualities are not the sole focus. Geo/graphic design is not simply a 'mode of telling', it is also a 'method of inquiry' (Richardson & Adams St. Pierre 2005).

Discussion

In conversation with participants about their mementos and collections various themes emerged. For example, it became clear that these objects enabled people to travel through both space and time. Objects brought them back to particular moments, places or people and, in doing so, triggered a range of emotions and sensory memories. The objects that enabled these imaginary and emotional journeys varied in both size and original manufactured 'purpose'. Ornaments are obviously designed to be displayed in a way that foregrounds their aesthetic qualities, but in this context, the aesthetic concerns recede and it is the ability of the material object to reconnect participants with their past that is important.

> The blue ceramic cat was given to me by the mother-in-law of a friend of mine who lived in France. I've had it with me in various homes on mantelpieces and it just reminds me of those times. I was staying on holiday and the Sunday habit was to go and visit the mother-in-law and have madeleines and aperitifs, Pernod or something of that sort, in her garden. She was an antique collector, the house was filled with all sorts of wonderful things and it was such a. . . such a 'French' afternoon and quite relaxed and elegant in a kind of way. I was just admiring some of the things in the house and she said would you like this. I demurred and then she insisted, and it became clear that it would hurt her feelings, that it would be even worse not to accept it than to accept it, so I did. It's followed me around, that was oh, a long time ago now, it would be the late seventies I suppose, seventy-eight, seventy-nine. It reminds me of being very young, it was quite exciting to go to Europe and be in France, and I guess it reminds me of that time too when I guess my life was a little less structured than it is nowadays.
>
> (JF 2009) (see Figure 6.13)

This cat, which wasn't deliberately chosen as a memento, and is described as 'following its owner around', has grown to be part of the participant's autotopography,

Figure 6.13 JF's blue ceramic cat

even though it seems she doesn't particularly care for its aesthetic qualities. Much like a stray cat who turns up so often it gets adopted, this cat has stood the test of time. It now provides a window to a time in the participant's life that was freer and filled with new adventures. The memories described are also multi-sensory – the tastes, smells, temperature, light and atmosphere are clearly conjured up by simply looking at it as it sits on her mantelpiece in London 30 years later. In this way, the memories clearly become an embodied, affective experience.

> *You talk quite specifically about the afternoon, and the food and drink. Does it literally almost conjure up the smells and the tastes and sounds?*
> Yes.
> *It's as though the object allows you to rewind time, it just transports you back to a particular moment.*
> Yes, it was a mild summer day, a July summer day, but Lille's climate is very similar to London, so it does not get hot, hot, hot summer days, but it can also have those mildish, gentle days with sun, but not very bright sun and that's the kind of day it was.
>
> (JF and author in conversation 2009)

Conversely, rather than being from the time or place that is being remembered, some objects are chosen specifically because there is something about them that makes a cognitive link to a particularly memory we are keen to carry with us.

> Penguins remind me of my Mum. It immediately made me think of her when I spotted it between old Oxo tins in a shop in East Dulwich. When I was a kid I used to call my Mum a penguin because I thought it was funny and cheeky, but it has grown to be an association of pure endearment. I still call her pengy Mum from time to time.
>
> (SG 2009)

In this context, the penguin could be seen as an empty receptacle that has been filled with memories upon purchase, rather than an object that already contains the memories on acquiring it. This response was from a participant in her mid-twenties who had moved to London from Liverpool, so it is likely she was furnishing one of her first homes away from home and was seeking to bring something of her previous life into her current one. It would also seem that this desire to reminisce and need to surround ourselves with meaningful objects that enable a nostalgic return to our past is not something that develops solely in later life.

Connections to friends and family, such as the penguin provides, were repeatedly cited in relation to a range of items.

> The horse calendar belonged to an elderly man who lived on a road I used to live on. Three siblings, all old East end costermongers from a large family of thirteen. Two of them were married and the five of them lived in this big old house on the corner, four doors down, and they were just so kind and supportive. None of them had any children and they kind of were the grandparents for the whole street. This used to be on his table in his kitchen, we used to go in for cups of tea. He and his brothers and the younger ones in the family used

to have a card school on Friday nights, and we would always go in afterwards and have a drink with them all, and find out who had won and that was always sitting there on the table.

(JF 2009) (see Figure 6.9)

Once again, the horse calendar hasn't been chosen for its aesthetic qualities, but specifically because it is full of memories from the room in which it resided in its previous life. The style of the ornament is in some ways the antithesis of the rest of the house; it is seemingly dated and out of place if taken at face value. This is often the way with our autotopographies, we don't choose them all at once and sometimes we don't even choose some of them ourselves. Therefore they become an eclectic mix that doesn't necessarily conform to current interior design trends. However, the point of these objects is what lies beyond their 'surface', and the horse calendar signifies a time when its current owner had moved to a different country and was taken in by a family that provided the warmth, love, care and support she wasn't able to receive in person from her own relatives.

Whilst ornaments are relatively small, larger items that might be considered as more 'functional' were also referred to. Forms of seating in particular were discussed by two participants as being evocative of particular memories.

This sofa was in their basement which was their sitting room and I can remember having just the most wonderful times sitting there laughing with them all and various other people on the street and you'd go in and Jimmy would be pouring drinks. Just really lovely times. For me, I'm an immigrant to this country. I'm came here in 1969, I don't have family here, I didn't have children and, I don't know, it's kind of like surrogate parents I suppose, and for them I think we represented the children they never had.

(JF 2009)

This chair makes me think of falling in love with London.

(SG 2009) (see Figure 6.5)

Chairs are often discussed in the context of memories. In *Ordinary Lives*, Highmore (2011) devotes several pages to a discussion of his 1970s Habitat chair and how it has been present for so many moments in his life and even now, as it sags with age, it remains a comfortable and comforting presence in his home. We sit in chairs, they literally hold us and bear our weight; similarly, we sink into a sofa, and its cushions mould to our bodies. Perhaps it is the act of sitting that creates this embodied connection with chairs and sofas that means they are likely to become a part of our 'physical map of memory, history and belief' (González 1995: 134).

Whilst autotopographies develop over time in something of a haphazard way, collections, by their very definition are developed with more intent. However, often collections can be sparked by a gift from someone else, or a purchase that was driven by some other reason than desire for the item. This is the case with CL's pigs as we see below.

> I wouldn't say that I'm an . . . I'm not an inveterate collector of one single item you know . . . the pigs came about by accident in a way, the paper-weights also. There was this very artistic friend who opened a shop on the Southbank, Gabriel's Wharf. So having been invited to the opening you sort of feel obliged to buy something don't you, so that's why I bought the pigs. Most of the other pigs have been given to me by my daughter.
>
> (CL 2009) (see Figure 6.14)

Some collections are almost the antithesis to Belk's (1995: 67) concept of 'actively, selectively, and passionately acquiring and possessing things'.

> The toilet roll holders, I was going to make and sell them. I think I was prob-ably feeling hard up at the time and I cast around to try and make a few things to sell and they were very easy. I made mice, dressed mice, jolly nice job, but my eyesight, I can't see to sew now really, then I made patchwork balls for babies, I've got several of those left.
>
> (CL 2009)

Figure 6.14 A small section of CL's collection of pigs

These collections are the result of failed attempts to develop hand crafted items for sale. Whilst the patchwork balls are not on display, and perhaps therefore aren't a collection in the sense that their owner still sees them as a failed attempt to produce a functional item with which to make some money, the toilet roll holders blur this line. They were made for the same reason as the patchwork balls and the mice, but they have graduated from being a product to sitting as a group within a section of the shelving that holds all the other collections on display. In this way the toilet roll holders have been transformed from 'marketplace origins to become a personal treasure' and as such have been 'recontextualised and elevated to a place of reverence' (Belk 2006: 541).

'Stocks', as CL described them in her original response to the postcard in the cultural probe, provide us with a further subset of collecting and collections. Stocks are stuff that has been collected with a view to future usefulness. There is a sense here that we are stocking up in order to be prepared for some unforeseen disaster. Then again, a world shortage of jam jars or a rash of buttonless cardigans are strange and unlikely disasters, even though they are being prepared for in many homes. This compulsion to hang onto things because 'they might come in handy someday' can become a 'grounding' for our identity that reduces the fear that we can somehow be erased ourselves (Belk 1988: 159). For some the stocks are piled through a desire not to throw away rather than to actively keep; for others, the reverse is true. Perhaps this type of collecting is driven by a mentality that comes with a certain age and experience. Those in the UK in the Second World War were encouraged to 'Dig for Victory' and Make Do and Mend' whilst rationing was in force. In today's throwaway society with planned obsolescence and fast fashion we are encouraged by many to dispose of stuff rather than keep it. Yet, neither do stocks sit particularly comfortably with the sustainability agenda – whilst they conform to the idea of not wasting things, they are also driven by a deliberate desire to increase their number, so align more with ideas of increased consumption rather than reduction. This conundrum and the guilt that may come with it is evident in the following exchange.

> I even bought some more arty materials on Thursday last week.
> *What was purchased last Thursday?*
> Well it's to do with this rubber stamping business.
> *Oh that's still ongoing is it?*
> Well I haven't done any for ages, but I saw this. . .
> (CL and author in conversation 2009)

The choice of words, or the lack of them is telling; 'I "even" bought some more' implies an incredulity that CL is still filling her house with more 'stuff'. The use of the word 'business' also implies a shorthand for something we're not really going to discuss in full, in the same way it might be used in the context of 'a messy business' or 'nasty business'. Finally, whilst admitting she hasn't actually used any of the rubber stamps for a while, her sentence trails away in an apologetic fashion. So in CL's case there also seems to be a guilt attached to her accumulation of stuff, and while she may call them 'stocks' and they don't conform to ideas of collecting in Belk's (1995) definition of the term, there is clearly a compulsion here.

Yet whilst there is a compulsion to stockpile that seems impossible for CL to resist, there is also an awareness of the accumulation of 'stuff' that will become someone else's problem after she dies. As she so aptly put in her original cultural probe response, 'what will my children do with all of it when I peg out?' CL's awareness of this situation has led her to begin to compile an 'artefacts list' that will be left for her children.

> I have almost finished what I call my artefacts list which is good because . . . whether my children will ever read it I don't know, but at least they've got the opportunity. So they will know where things have come from.
> *I guess that's almost like a guidebook . . .*
> Yes it is.
> *. . . if you are going to go back to that analogy of the things being a landscape, it's almost like a Wainwright's walk to . . .*
> Yes, yes, through my life. Yes.
>
> (CL and author in conversation 2009)

CL is therefore positioning herself not as an owner of this stuff, but as a caretaker. Her collecting is preserving this stuff – and some of the items are family heirlooms – for future generations, though equally she is painfully aware her children may not be interested. Yet, the listing is not just for them, subconsciously it is undoubtedly part of the memorialising process. Through it CL is literally able to take stock, to reflect on her life and map out its journey using the co-ordinates provided by her growing collection of stuff.

Summary

The idea of home does not just relate to a house or bricks and mortar, it is more than that. Home is also 'a set of feelings' and 'the relations between this material and affective space' (Blunt & Dowling 2006: 22). One of the ways we transform a house into a home is through our 'stuff'. Our material possessions aren't simply functional, many of them – particularly mementos or souvenirs – have a deep emotional connection. Our 'stuff' therefore acts as material prompts to the narrative of our lives, forming a 'spatial representation of identity' – an 'autotopography' (González 1995: 133). This enables us to bring the past into the present and 'travel' both temporally and spatially. Yet, for those who don't own these particular possessions they can seem like a jumble of inconsequential, meaningless items – their agency is diminished, they become inert.

Stuff reflects three participants' responses to the question 'what makes your house a home?' that was included within a cultural probe pack. It draws together a participant's life story, an academic essay and conversations about 'stuff', as well as documentary photography. The experimental book challenges Ingold's perception of the silent printed page and engages the reader in the production of meaning. The design interventions that enable this work with both the typographic content and the structure and form of the book. Reading *Stuff* and handling the evocative items contained within it enables an individual, personal journey that provokes one's own memories and emotions. However, the process of the project also acts as a 'mode of inquiry', going beyond a simple 'mode of telling' (Richardson & Adams St. Pierre 2005: 923). Thus the process of the project is able to provide a range of insights, for example, how such objects might be chosen or acquired; what kind of memories they might trigger; whether their aesthetic qualities are important to their owner; whether stocks are driven by ideas of sustainability or an urge to consume; and that we are all only ever caretakers of our stuff, that someday someone else will either inherit it or have to dispose of it on our death. In this way the livre d'avant garde does challenge both art and life in that it offers both a re/presentation of place, but also understanding. Whilst not the format of a traditional academic book *Stuff* does offer a way of addressing Pink's concerns that this type of multi-sensory, embodied experience is difficult to express or capture within responses that are text based (Pink 2015: 40).

Much like the previous project, *Stuff* was developed over a long period of time, engaging in sustained ethnographic research and involving participants. However, sometimes we visit particular places that make us feel a certain way – we can sense atmospheres, or feel the spectral presence of previous inhabitants. The following chapter centres on this affective experience of place, and therefore explores the potential of a geo/graphic approach in re/presenting embodied experiences of place that 'we cannot explain, do not expect, understand, or struggle to represent' (Maddern & Adey 2008: 292).

Bibliography

Arnar, A. S. (2011) *The Book as Instrument: Stephane Mallarme, the Artist's Book, and the Transformation of Print Culture*. Chicago: University of Chicago Press.

Belk, R. W. (2006) 'Collectors and Collecting', in Tilley, C., Keane, W., Küchler, S., Rowlands, M. & Spyer, P. (eds) *Handbook of Material Culture*. London: Sage, pp. 534–545.

Belk, R. W. (1995) *Collecting in a Consumer Society*. London: Routledge.

Belk, R. W. (1988) 'Possessions and the extended self', *Journal of Consumer Culture*, 15(2), pp. 139–168.

Bhatti, M., Church, A., Claremont, A. & Stenner, P. (2009) 'I love being in the garden': enchanting encounters in everyday life', *Social & Cultural Geography*. 10(1), pp. 61–76.

Blunt, A. & Dowling, R. (2006) *Home*. Abingdon: Routledge.

Boehner, K., Vertesi, J., Sengers, P. & Dourish, P. (2007) 'How HCI interprets the probes', *Proceedings of the SIGCHI Conference on Human Factors in Computing Systems*, San Jose: USA, 28 April to 3 May, 2007, pp. 1077–1086.

Borges, J. L. (1977) *The Gold of the Tigers: Selected Later Poems: A Bilingual Edition*. Translated by Alastair Reid. New York: E. P. Dutton.

Boym, S. (2007) 'Nostalgia and its discontents', *The Hedgehog Review*. 9(2), pp. 7–18.

Brickell, K. (2012) '"Mapping" and "doing" critical geographies of home', *Progress in Human Geography*. 36(2), pp. 225–244.

Cardinal, R. (2001) 'The Eloquence of objects', in *Collectors: Expressions of Self and Other*. Shelton, A. (ed.) London: Horniman Museum and Gardens, pp. 23–31.

Carrion, U. (2001) *The New Art of Making Books*. Nicosea: Aegean Editions.

Coffey, A. & Atkinson, P. (1996) *Making Sense of Qualitative Data: Complimentary Research Strategies*. London: Sage.

Crang, M. & Cook, I. (2007) *Doing Ethnographies*. London: Sage.

Csikszentmihayli, M. & Rochberg-Halton, E. (1981) *The Meaning of Things: Domestic Symbols and the Self*. Cambridge: Cambridge University.

Dasté, O. (2011) 'The suitcase' in Turkle, S. (ed.) *Evocative Objects: Things We Think With*, Cambridge, Mass.: MIT Press, pp. 245–249.

Del Casino, V. J. & Hanna, S. P. (2006) 'Beyond the "binaries": A methodological intervention for interrogating maps as representational practices', *ACME: An International Journal for Critical Geographies*. 4(1), pp. 34–56.

Dowling, R. & Power. E. R. (2013) 'Domesticities', in Johnson, N. C., Schein, R. H. & Winders, J. (eds) *The Wiley-Blackwell Companion to Cultural Geography*. Chichester: John Wiley & Sons, pp. 290–304.

Drucker, J. (2004) *The Century of Artists' Books*. 2nd edn. New York: Granary

Drucker, J. (2003) *The Virtual Codex from Page Space to E-space*. Available at: www.philobiblon.com/drucker/ (Accessed 8 December 2017).

Eco, U. (1986) 'Architecture and memory'. *Via*. 8, pp. 88–94.

Gaver, B., Boucher, A., Pennington, S. & Walker, B. (2004) 'Cultural probes and the value of uncertainty', *Interactions*. 11(5), pp. 53–56.

Gaver, B., Dunne, T. & Pacenti, E. (1999) 'Cultural probes', *Interactions*. 6(1), pp. 21–29.

González, J. A. (1995) 'Autotopographies', in Brahm, G. & Driscoll, M. (eds) *Prosthetic Territories: Politics and Hypertechnologies*. Boulder: Westview Press, pp. 133–150.

Hecht, A. (2001) 'Home sweet home: Tangible memories of an uprooted childhood', in Miller, D (ed.) *Home Possessions: Material Culture Behind Closed Doors*, Oxford: Berg, pp. 123–145.

Hemmings, T., Clarke, K., Rouncefield, A., Crabtree, A. & Rodden, T. (2002) 'Probing the probes', *Proceedings of the Participatory Design Conference*, Malmo: Sweden, 23 to 25 June 2002, pp. 42–50.

Highmore, B. (2011) *Ordinary Lives: Studies in the Everyday*. Abingdon: Routledge.

Highmore, B. (2002) *Everyday Life and Cultural Theory: An Introduction*. London: Routledge.

Hochuli, J. (1996) *Designing Books: Practice and Theory*. London: Hyphen Press.

Hockey, J., Penhale, B., & Sibley, D. (2005) 'Environments of memory: Home space, later life and grief', in Davidson, J., Bondi, L. & Smith, M. (eds) *Emotional geographies*. Aldershot: Ashgate, pp. 135–146.

Ingold, T. (2007) *Lines: A Brief History*. Abingdon: Routledge

Jääsko, V. & Mattelmäki, T. (2003) 'Observing and probing', *Proceedings of the 2003 International Conference on Designing Pleasurable Products and Interfaces*, Pittsburgh: USA, 23 to 26 June 2003, pp. 126–131.

Kitchin, R. & Dodge, M. (2007) 'Rethinking maps', *Progress in Human Geography*. 31(3), pp. 331–344.

Lorimer, H. (2008) 'Poetry and place: The shape of world', *Geography*, 93(3), pp. 181–182.

Maddern, J. F. & Adey, P. (2008) 'Editorial: Spectro-geographies', *cultural geographies*. 15(3), p. 291–295.

Mau, B. & Mermoz, G. (2004) 'Beyond looking: Towards reading', *Baseline*. 43, pp. 33–36.

Meah, A. & Jackson, P. (2016) 'Re-imagining the kitchen as a site of memory', *Social & Cultural Geography*. 17(4), pp. 511–532.

Niles, J. D. (1999), *Homo Narrans: The Poetics and Anthropology of Oral Literature*. Philadelphia: University of Pennsylvania Press.

Patton, M. (2002) *Qualitative Research and Evaluation Methods*. London: Sage.

Pearce, S. M. (1994) 'Objects as meanings; or narrating the past', in Pearce, S. M. (ed.) *Interpreting Objects and Collections*. London: Routledge, pp. 19–29.

Pink, S. (2015) *Doing Sensory Ethnography*. 2nd edn. London: Sage.

Pollack, S. (2011) 'The rolling pin' in Turkle, S. (ed.) *Evocative Objects: Things We Think With*. Cambridge, Mass.: MIT Press, pp. 225–231.

Proust, M. (2013) *In Search of Lost Time, Volume 1: Swann's Way*. New Haven: Yale University Press.

Rand, P. (1970) *Thoughts on Design*. 3rd edn. London: Studio Vista.

Relph, E. (1976) *Place and Placelessness*. London: Pion.

Richardson, L. & Adams St. Pierre, E. (2005) 'Writing: A method of inquiry', in Denzin, N. K. & Lincoln, Y. S. (eds) *Handbook of Qualitative Research*. Thousand Oaks: Sage, pp. 959–978.

Robertson, S. (2008)'Cultural probes in transmigrant research: A case study', *InterActions: UCLA Journal of Education and Information Studies*, 4(2) Art. 3, Available at: https:// escholarship.org/uc/item/1f68p0f8 (Accessed: 8 December 2017).

Rose, G. (2003) 'Family photographs and domestic spacings: A case study', *Transactions*. 28, pp. 5–18.

Rose, G. (1993) *Feminism and Geography: The Limits of Geographical Knowledge*. Cambridge: Polity Press.

Shelton, A. (2001) 'Introduction: The return of the subject', in *Collectors: Expressions of Self and Other*. Shelton, A. (ed.) London: Horniman Museum and Gardens, pp. 11–22.

Turkle, S. (ed.) (2011) *Evocative Objects: Things We Think With*. Cambridge, Mass.: MIT Press.

Tuan, Y-F (1980) 'The significance of the artifact', *Geographical Review*. 70(4), pp. 462–472.

7   An embodied, affective experience of place

Old Town

Introduction

The previous two chapters discuss projects that were developed over long periods of time, each drawing together content that variously includes ethnographic research, participant contributions or autoethnographic writing, images, and contextual research. However, this chapter features a project that focuses on a much more immediate, instinctive experience of place, and in doing so, it further engages with ideas of the more-than-representational, and embodied, affective encounters. The small experimental book (13.5 x 18.5cm) discussed – *Old Town* – was developed during the week-long *Experimenting with geography: See, hear, make, do* workshop held at the University of Edinburgh in 2010.

Inspired by the experience of wandering through the wynds, closes and courts of Edinburgh Old Town, the design explores the potential of the book to create an affective experience of place. It uses the graphic spaces of a facsimile edition of William Edgar's engraved 1765 map of the Old Town to recreate the physical spaces of light and dark that I experienced as I explored the area. Texts relating to characters, buildings and events from the eighteenth and nineteenth centuries are interspersed within the folds of the pages, adding a sense of the past to the immediate experience of the present.

Much like any city, Edinburgh reveals fragments of its past in ways that span formally designed, planned and marketed heritage sites through to material traces that are the product of the passage of time and people. These palimpsestual traces can be described as spectral presences that inextricably link the immaterial with the material and reveal traces of absence within the present. Although these absences may trigger memories, particularly for those who have a previous knowledge of the place, by nature the narratives of these absences are fragmented and non-linear. Therefore, new ways of engaging with these narratives and with the more-than-representational aspects of memory are required ' . . . [M]emory demands new ways of writing; narratives that better cope with our fragile and contingent recollections, disclosing the haunting presence-absence of the spectral in all its shapes, apparitions and phantasms' (Hill 2013: 379).

Old Town therefore maximises the potential inherent in the physical and material form of the book. It offers a reading experience that engages with the paradoxical nature of absent presence and the inevitably partial narratives that emerge from this version of place. Unlike the previous books, the reader's engagement is primarily experiential, thus the book is developed using no traditional content, therefore the analysis undertaken during the design development is primarily materials led. However, in undertaking the initial ethnographic research, the Old Town was explored on foot alone and with walking tour groups; the Real Mary King's Close heritage centre was visited; and documentary photography was undertaken where possible. From this preliminary research, insights are thus available as to how Edinburgh engages with its past within the present and how it packages and promotes particular aspects of its story for consumption by tourists. Whilst this might not be evident to the readers of the book, it is important in the context of the geo/graphic design process. There is a danger with research that could be considered experimental or experiential that it privileges one type of work over another and reiterates a divide between theory and practice (Nash 2000). However, with the inclusion of historical texts within the book, the associated analysis of the process of its making, and the analysis of the ethnographic research, a more holistic approach is developed, one that closes this divide.

Edinburgh

The Scottish capital since the fifteenth century, Edinburgh is home to around half a million people and is situated on the Firth of Forth. It is built on a spectacular geological setting with the cityscape defined by its rocky natural landscape that was moulded by glaciers (McMillan & Hyslop 2008: 1). The city is also characterised by its tall, stone constructed buildings, giving it a sense of permanence, standing steadfast in the face of the cold wind, rain and snow that frequently assault the city in the cooler months. However, in the 1400s, Edinburgh buildings were predominantly made of timber, with thatched roofs, and were usually no more than two storeys high. Within the confines of the Old Town this posed an obvious fire risk and laws were passed in 1425 that further buildings should be erected using stone (McMillan & Hyslop 2008: 4). Edinburgh's cityscape is therefore dominated by buildings erected with locally quarried sandstone. The early quarries were situated just outside the town walls or near to the houses and lands within the walls that were being built; thus Edinburgh's 'stone-built heritage literally grows out of the bedrock foundations of the city' (McMillan & Hyslop 2008: 1).

Edinburgh's Old Town, along with the New Town, was designated a UNESCO world heritage site in 1995. The Old Town stretches along either side of the Royal Mile and High Street, that lead down from the Castle to Holyrood Palace. Much of the Old Town's original herring-bone street pattern has survived and is little altered since medieval times. The narrow closes, wynds and courts are flanked by tall tenement buildings, giving it a warren-like feel within the dark enclosed spaces that are formed by the height of the buildings, the lack of space between them, and the narrowness of the closes and wynds. The decision to build upwards rather than outwards was due to the lack of suitable land around the Old Town, and the desire to remain within the fortified walls of the existing city and close to the castle defences. This led to squalid conditions: 'Prone to fires, repeatedly ravaged by disease, it was notorious for its drunkenness, malevolent ghosts and violent crime. The 50,000 residents and freely wandering livestock were trapped in its narrow streets, defensive walls, steep ravines and tottering tenements' (Campbell 2016: no pagination). The steep ravines acted as open sewers for the disposal of both human and non-human waste, and as we will see later, stories of this type of squalor have become key to the telling of Edinburgh's past in many contexts. However, present day Edinburgh is somewhat more sanitary, and the closes and wynds can be wandered without fear of encountering the plague – and many people do come to Edinburgh to do just that. Tourism is big business, with £1.32 billion being spent in 2015 by 3.85 million visitors, according to the Edinburgh Tourism Action Group (ETAG 2016: 7). Of these visitors, 77% chose it as a destination due to its historic city centre, and 91% of visitors listed 'walking around the city' as one of their top ten attractions (ETAG 2016: 10–11). The Old Town is full of pubs, restaurants, independent shops, galleries and arts venues, and retains many sixteenth and seventeenth century buildings. Its UNESCO listing focuses particularly on the fact that its level of 'authenticity' is high and that 'the high-quality buildings of all dates have been conserved to a high standard and the layout of streets and squares maintain[s] their intactness' (*Old and New Towns of Edinburgh*, no date). Because of this, visitors are able to feel they are 'stepping back in time', walking the same streets and following in the same footsteps of the famous – and infamous – Edinburgh residents that have preceded them.

Materiality, absence and presence

The city is often discussed in terms of a palimpsest – a piece of medieval writing material – usually parchment – which was reused time and time again, with each inscription being erased and another written over it. The earlier inscriptions would never quite be totally erased, so the result was a build-up of the sum of all the inscriptions. Similarly, 'cities are archives of the way things used to be' (Dittmer 2014: 482) and the processes of social, economic and political change see the continual loop of 'demolition, replacement, renovation and reconstruction of buildings' that produce this material and temporal 'collage' that is evident in the palimpsestual landscape (Edensor 2012: 448). In amongst this ongoing development, cities throughout the world construct particular buildings and sites in order to remember, to celebrate and to commemorate. War memorials, cemeteries, museums, and temples are all 'storage houses of former times', thus architecture's remit is not simply to design the urban landscape and order space, it also lies with the 'preservation of time' (Quiring 2010: 199). These formal monuments position memory within the realm of social and political processes and are an attempt to 'fix authoritative meanings about the past'. Often focusing on militarised, 'masculinised, classed and racialised ideologies', they are also often contested and controversial (Edensor 2005a: 830). Similarly, formally managed and designed heritage sites within cities – which might include whole areas or single buildings – also provide visitors with a fixed narrative that limits their 'interpretive and performative scope' and banishes the 'ambiguity and multiplicity' of the past (Edensor 2005a: 831). However, although such sites do attempt to fix the past, there remains 'a kind of memory surplus', particularly with material that is 'informal, vernacular and sensual', that can escape the constraints of such commodified spaces (Moran 2004: 58). This idea of a memory surplus is perhaps partly related to the distinction between practices of 'inscription' and 'incorporation'. Inscription relates to the representation of memory, of sites that contain information, unlike incorporation which relates to the embodied act of remembering (Hill 2013: 380). There is a further link here to ideas of voluntary and involuntary memory as discussed in the previous chapter, with formal, inscribed sites of memory prompting voluntary responses, yet still retaining the potential for an aspect of them to provoke an involuntary memory for some visitors. In some cases, such involuntary memories may well be triggered for those who have prior experience of the site or artefact before it was incorporated into a heritage site, but for others, they remain invisible (Meier 2013: 475).

For Nora (1989), such formal monuments and buildings are designed to organise a past that will inevitably be forgotten due to the drive for change. We therefore design these sites of memory, or 'lieux de mémoire', 'because there are no longer milieux de mémoire, real environments of memory' (Nora 1989: 7). However, it's not just formal institutions such as these that preserve time; traces of the past can be found everywhere we look. In worn stone steps, masons' marks and graffiti, for example, such physical reminders still exist (and continue to be produced) across the city. In major cities across the world, forms of stone, brick or concrete are used to construct the majority of buildings, and as we have seen above, Edinburgh is no different in this respect. A 'stony materiality' such as Edinburgh's facilitates an 'affective, sensual and imaginative engagement' that evokes the absence of 'distant lives and events, human remains, cultural practices and tastes'. These absences are therefore 'made present and acknowledged' through the material form of the buildings themselves (Edensor 2012: 447). The stone that forms many of Edinburgh's buildings is particularly enduring, and it is partly this endurance and the intact nature of the city that therefore 'offers a means of experiencing and thereby presencing past events' (Hill 2013: 381)

Spectral presences

Spectrality has emerged as a theme within geography since the late 1990s (Meier 2013: 470) and the 'figure of the ghost is often used as a means of apprehending that which we cannot explain, do not expect, or understand, or that we struggle to represent' (Maddern & Adey 2008: 292). Such ghostly presences may seem far-fetched initially, but they are everywhere if we look – to the extent that de Certeau claims 'haunted places are the only ones people can live in' (1984: 108). As we discussed in Chapter 1, place is always in process (Massey 1994, 2005) and is continually reconstructed via both human and non-human activity. Yet, whilst buildings are demolished, atmospheres change, and the flow of people might shift accordingly, the previous structures are imbued with a resilience. Thus 'this durability of meaning and of materialities' makes visible 'the transformation of what was there before and is now absent' (Meier 2013: 475). Ideas of haunting thus bring the past into the present, and in a sense fold time, unsettling the linearity we are used to (Meier 2013: 470). The spectral thus positions the past and the future as co-existent and interacting in 'uncertain and unpredictable ways' (Hill 2013: 381): '[M]emory is born of strange and uncanny associations, inexplicable connections between times and places that erupt into the present without warning '(Hill 2013: 379). Hauntings involve the 'just perceptible, the barely there, the nagging presence of absence', opening our eyes to 'the persistence of presences that somehow remain' (Maddern & Adey 2008: 292–293). Such ghostly presences that produce echoes of the past also have 'the power to move us in unexpected ways' – they can be affective, evocative and unsettling (Hill 2013: 380).

Time is not only preserved through the physical, material traces evident within the fabric of buildings, it is also apparent in the presence of buildings from different eras, or the empty spaces left where buildings have been demolished or destroyed. Both the material traces and these absences often provoke an 'empathetic conjecture' rather than an awareness of specific loss (Edensor 2012: 448) – though as we have seen previously, in Chapter 2, long term residents of a city will be aware of landmarks that no longer exist and for them this absence may provoke a more specific emotional response and associated memory. This patchwork construction reveals multiple temporalities within the city which combine to produce an 'asynchronous mélange, a play of temporal juxtapositions that incites an improvisational and fragmented account rather than a sequential narrative' (Edensor 2012: 450). These 'traces . . . pervade the mundane and spectacular spaces of the city', interrupting 'the flow of the present, interjecting with inferences, affects, sensations and fantasies' (Edensor 2012: 448). If we 'become attuned' to these traces – through historical research and developing our local knowledge – our interaction with the materiality of the city enables an intimate connection with 'distant presences, events, people and things'. However, this 'attunement' is also triggered by our embodied, affective encounters and the resulting empathetic response that might be described as a 'peculiar, inarticulable feeling of pathos' (Moran 2004: 58). Unlike the heritage industry which packages the narratives it purveys, memories triggered by everyday experiences of the material environment of the city resist this organised version of nostalgia. As discussed in the previous chapter, this type of memory is involuntary, emerging unexpectedly, and the resultant memories and feelings such everyday encounters trigger can be unsettling, forcing 'us to contemplate the transience of unacknowledged lives' and evoking a 'specific kind of sadness' (Moran 2004: 61).

This therefore suggests that the immaterial is inextricably linked with the material. Indeed, absence can be seen as a 'relational phenomenon, something that is produced in the back and forth between absence and presence, materiality and immateriality, the social and the natural' (Meier, Frers & Sigvardsdotter 2013: 424).

Thus, absence is not a 'thing' that exists itself, rather it is the relations between these states that bring the absence into being (Meyer 2012: 107). Absence is therefore something that is 'performed, textured and materialised through relations and processes, and via objects'. It can therefore be traced and is perhaps best thought of as a trace, rather than as a connection 'between the absent and the present' (Meyer 2012: 107). For Meyer, the word trace is particularly apt as its meaning extends beyond that of the previously discussed physical trace in the environment. Trace also means to locate or follow, and it denotes 'an active and spatial act of mapping out', and refers to 'something that is incomplete, something that once was' (2012: 107). According to Meyer (2012: 107), to trace absence is therefore to undertake a performative and spatial act, yet much like the discussions relating to non-representational theory and text-based versions of events (Dirksmeier & Helbrecht 2008) in Chapter 2, this tracing is unable to capture the absence fully as it is 'always following and always behind'. Absence is thus always incomplete and therefore resists definition; it is, by its very nature 'absence', and is therefore always unattainable, never present (Meyer 2012: 107).

As we have seen above, absences can emerge and interrupt the present through material objects and our associated memories, thoughts and emotions, and absences can also emerge through 'talk and texts'. Through this combination of objects, practices and feelings, it is evident that absence has agency and can therefore affect the social world – 'In other words, as well as being something we engage with, absence *does* things' (Meyer 2012: 104; italics in original). Yet, whilst absence is a not a 'thing' in itself, Meyer is less clear as to what 'gives life' to absences (Meier, Frers & Sigvardsdotter 2013: 424). For Meier, Frers & Sigvardsdotter, these absences are experiences and, therefore, they emerge 'only, and without exception, in lived experience' (Meier, Frers & Sigvardsdotter 2013: 424). This lived experience is important in two ways. First, it is the embodied experience which enables place related memories to bring history to life. Second, it is through our prior lived experience that we lay down these memories which are ultimately as important as the material traces in bringing such absences to life (Meier, Frers & Sigvardsdotter 2013: 425).

As we have seen in the previous chapter, in the context of our own personal memories, places of importance to us, such as our home, are likely to trigger a narrative that recounts, in part at least, the history of our life. Long dead friends and family, momentous conversations, and pivotal moments remain vivid to us, along with many other moments that are more mundane in their telling. Our autotopographies (González 1995) ensure that the smallest ornament or torn bus ticket can conjure up memories that are as vast and important as the objects are conversely small and seemingly inconsequential. A single material possession from one's life therefore has the capacity to evoke memories that encapsulate a much bigger picture. In places that are less personal to us, a similar relationship between the fragment and the whole may still play out. For example, when the demolishing of an end terrace wall enables us to view the exposed interior of an abandoned family home, complete with faded wallpaper and cracked tiles around the hearth, we may begin to imagine the occupants, and in doing so, bring our own memories of such places to bear on the situation. In this context, 'the power of synecdoche in landscapes is that such a fragment takes on greater meaning: the projected meaning of the imagined whole' (DeLyser 2001: 27).

Ruins

In terms of 'fragments', much has also been written about ruins in relation to our sense of temporality and our relationship to the past. Ruins have long since been subjected to a romanticised gaze and ideas of the picturesque which conjured up melancholic associations with the past (Edensor 2005b: 323). This privileging of the visual – which could also be said to be the case for geography's encounters with architecture generally (Paterson 2011: 264) – and the desire to engage with the type of ruins that illustrate a romantic aesthetic, neglect the potential of 'the ruin's non-representational power to activate memory and sensation and downplay the significance of . . . lived presence' (DeSilvey & Edensor 2012: 467). Much writing about ruins focuses on our embodied engagement with them, and positions this in direct contrast to our experience of walking in the majority of cities within the West, which Edensor describes as 'increasingly desensualised' (Edensor 2008: 130). This 'smoothness of space' in the contemporary Western city regulates and restricts sensory experience (Edensor 2005b: 324). For example, smooth walkways and pavements, and easily identifiable kerbs, impact on how we physically sense the world as our 'feet and legs are not enlivened by contact with the ground' because there is no unpredictability or irregularity to negotiate (Edensor 2008: 131). Further sensory regulation is also present in the form of 'clear and linear sight lines, deodorised environments, [and] highly regulated soundscapes' (Edensor 2005b: 324). This attempted erosion and marginalisation of 'other kinds of sensory experience' in such 'purified' environments results in the foregrounding of a 'visual consumption' of place (Edensor 2008: 134). Similarly, in the places that commodify memories, such as heritage sites, space and objects are ordered so as to function as 'icons of memory, cultural or historical exemplars or aesthetic focal points (Edensor 2005b: 312). These type of actions thus effect 'a coding that "brands" space and reduces it to a few key themes' (Edensor 2008: 134). For Edensor, these types of sanitising strategies deliberately attempt to 'exorcise haunted places' in order to minimise their effect on the linear narrative that the site is attempting to construct. However, as we have seen above, it is difficult to contain memories in this way; those that are involuntary are likely to irrupt into our consciousness and avoid erasure (Edensor 2005a: 829).

In contrast, encountering ruined spaces that have avoided exorcism enlivens the body and challenges it through 'a wealth of multi-sensual effects – including smells, sounds and tactilities' – which therefore mitigates against a visual dominance and offers a range of 'rich and unfamiliar affordances' (Edensor 2008: 132). Thus the experience of ruins is an affective one, as the materiality of the ruin – its textures, smells, dampness, and uneven or broken surface – impact upon the body (Edensor 2005b: 324). We therefore have to be continually aware of our step, endeavouring not to trip or fall, avoiding touching jagged edges, or ingesting potentially harmful substances. The ruin is therefore positioned as the antithesis to the regulated space of a contemporary Western city or the designed and managed space of a heritage site. Rather than engage with a version of the past that is literally sanitised and managed, ruins therefore offer us the 'capacity for alternative, sensual engagements with the past' (DeSilvey & Edensor 2012: 471). Once again, unlike the linear narratives of the heritage site, an engagement with the past in a ruined space is one in which stories remain inevitably partial and speculative, unable to be 'composed into coherent temporal sequences' (Edensor 2008: 137).

> Ruins foreground the value of inarticulacy, for disparate fragments, juxtapositions, traces, involuntary memories, uncanny impressions, and peculiar atmospheres cannot be woven into an eloquent narrative. Stories can only be contingently assembled out of a jumble of disconnected things, occurrences, and sensations.
>
> (Edensor 2005a: 846)

We often build these 'contingent stories' through an interface of both personal and collective memory, as the material remains we encounter in ruins 'mediate between history and individual experience' (DeSilvey & Edensor 2012: 472). In such a material encounter, the ruin becomes an 'experiential' space— those who once used to live or work within the site are brought to life once again through our imagination which is stimulated by this 'embodied exchange' with history (Garrett 2011: 1057). Thus, ruins 'impose their materiality of the sensory experience of visitors' and in doing so they bring to life 'the forgotten ghosts of those who were consigned to the past' (Edensor 2005b: 311). So, rather than the memories remaining invisible to those who have no prior experience of the site (Meier 2013: 470–471), these absences make their presence felt by 'haunting the visitor with vague imitations of the past' and in doing so thus avoid fixity and 'haunt the desire to pin memory down in place' (Edensor 2005a: 829).

Whilst Edinburgh has little in the way of ruins such as might be described by those writing above, what it does have in the Old Town is a cityscape that is dominated by very old buildings, cobbled streets, dark, damp wynds and steep, stepped closes, the majority of which are so narrow they are without names, not even present on Google maps. Therefore, even though the city has been designated a UNESCO world heritage site, and the city certainly has many buildings and tours that attempt to reify memory, it is impossible not to encounter a similarly sensorial experience with the material form of the Old Town and therefore one does engage in an embodied exchange with history. I would also argue that many cities, Edinburgh included, are as confused and confusing as ruins, and resist a linear narrative. Their patchwork development, as Edensor (2012: 450) suggests himself, gives rise to asynchronous temporalities and narratives that are fragmented, so there exists some rich middle ground between a crumbling industrial ruin and a designed and managed heritage site. Also, unlike ruins, whose ghosts are said to be mundane, many of the ghosts that are resident within Edinburgh's Old Town are more melodramatic, often 'emanating from dire trauma or atrocities' to 'strike terror into the haunted' (Edensor 2005a: 836). Yet this is not the case for all of them. Amongst the repeatedly summoned up spectres of infamous Edinburgh residents, other more 'indeterminate absent presences' are evident within the walls of the Old Town.

A geo/graphic approach to Edinburgh Old Town

During 2010 I was offered an opportunity to participate in the week long, Economic and Social Research Council (ESRC) funded, *Experimenting with geography: See, hear, make and do* (EWG) workshop at Edinburgh University. The workshop was developed in order to explore the idea that social research could benefit from the adoption of some of the techniques used within different types of creative practice and it brought together a range of early career researchers and doctoral students, from both the social sciences and art and design. Although many of the participants were engaged in sound and video work, and therefore much of the focus of the week was on those areas, there was an experiential immediacy to many of the approaches that inspired me to explore place in a different way and seek out the potential residing in more 'immediate', embodied experiences. As a designer I was taught to research the brief thoroughly in a relatively traditional way, and to develop a range of ideas that are capable of being defended or reinforced by that research. In Edinburgh I was surrounded by sound, video and performance artists, and one in particular was running drumsticks along railings just to see what it sounded like, jumping onto a bike to film herself cycling to a derelict Wild West town south of the city, and recording howling dogs to create a canine reworking of Ennio Morricone's spaghetti Western soundtrack. Being literally out of my usual place, and surrounded by people working in a different way enabled me to shift my thinking and approach slightly. This approach, primarily using walking again, perhaps wasn't as extreme as those listed above, but it became more 'hands on' – foregrounding an embodied experience of place – and consequently, although somewhat paradoxically, was a way of letting go more. This is not to suggest that I was somehow sitting on the fence, adopting neither a hands-on nor a hands-off approach. Rather, it was a realisation that I could complement what might be considered normative graphic design research strategies with more immediate, experiential understandings of place (see Barnes 2012).

Only in Edinburgh for a week I had a limited time both to explore place and to develop my ideas. It was a place I had visited once before, but not for many years, so I was arriving without many preconceptions. As I arrived at Waverley Station I was immediately struck by the solidity of the city. It seemed to be literally hewn from the rock, with the stone facades facing the elements, silently enduring season after season, year after year. I found myself running my hands across the stone as I walked, feeling textures made by nature, by stone carvers and by the traces of 500 years of occupants passing through the city. I became fascinated by the closes, wynds and courts of the Old Town and as one walks along the Royal Mile, the entrances of these alleyways beckon the visitor in, giving glimpses of small courtyards or steps that suddenly transport one, as if by a magical

Figure 7.1 A narrow close within Edinburgh Old Town

shortcut, to a lower part of town (see Figure 7.1). When exploring these spaces one constantly moves from dark to light and back again – from being enclosed by walls and a low roof, to standing under the open sky within a courtyard.

Figure 7.2 Plaque at the entrance of Lady Stair's Close

Thus temperatures change from a chilly, damp, dark space to a warmer sunlit one, and smells congregate within the narrow confines of the low ceilinged closes only to dissipate once in the open air of the courtyards. The history seems tangible within the walls themselves, but is also overtly recorded in a series of plaques relating to names such as Fleshmarket Close or Lady Stair's Close, and residents of note (see Figure 7.2). It was these embodied experiences of this disorienting, affective place that I wanted to communicate. Initially, therefore, I spent several days just walking the Old Town, recording my thoughts and experiences and photographing formal traces of the past such as signage and plaques and informal physical traces of wear or use evident upon the brick and stone. As part of my explorations, I also took part in two free walking tours – one history tour and one ghost tour. I also visited The Real Mary King's Close, a heritage site and tourist attraction that offers tours of intact seventeenth century streets that were built over and are now situated under the current City Chambers site. Whilst these more formally commodified experiences of the Old Town proved less useful for the development and production of the book itself, they provided further insights into how tourism within Edinburgh capitalises on the spectral traces evident in the material and sensory experience of the Old Town, and these will be discussed in more detail below.

Unlike the project discussed in Chapter 5 that explores space in a relatively systematic way, there was no system at play here. My wanderings were much more dérive like, I simply entered closes and wynds at will, just to see where they might lead and what I might find – I wasn't concerned about covering the whole of the Old Town or developing a mental map of every single route in and out of it, I was simply interested in experiencing it. These tactics and the process of the project bring to mind several of Bruce Mau's (Maclear & Testa 2000: 88–91) statements from his *Incomplete Manifesto for Growth*, for example:

> *Process is more important than outcome*
> When the outcome drives the process we will only ever go to where we've already been. If process drives outcome we may not know where we're going, but we will know we want to be there.

> *Drift*
> Allow yourself to wander aimlessly. Explore adjacencies. Lack judgment. Postpone criticism.

Engaging with place in this way seems appropriate, as it is always ongoing, always undetermined and always unfinished (Massey 2005: 107). Therefore, there is no single totalising 'truth' about place waiting to be discovered, and narratives of place are multiple, contingent and fragmented. However, because of this there inevitably remains the challenge of constructing 'a narrative of the past that attends to the discursive and embodied conditions of its existence in the present, to highlight the matter that brings forth memories, and to make visible the invisibility of the spectral' (Hill 2013: 382).

Re/presenting place

A moment of chance shifted my thinking about the visible form of the project very quickly. As I was walking down the Royal Mile I noticed several facsimile editions of old maps of Edinburgh displayed in the window of a shop. William Edgar's map of the Old Town from 1765 had originally been etched and the style of the cross hatching used to show buildings created a dramatic distinction of negative and positive space on the map. I immediately began to think of these negative and positive spaces in relation to the light and dark of my Old Town experience. In trying to recreate the embodied, affective dimension of exploring these physical spaces within the pages of a book, the facsimile map thus became a way to place the reader 'within' the space rather than above it. I wanted to emphasise the disorienting nature of the Old Town and the changes from light to dark, and back again. To this end, I scanned the map and began to enlarge sections of it, which I printed onto A4 paper. As I enlarged the images, they retained a sense of 'mapness', but also took on a more abstract geometric sense of negative and positive space. As I did this I realised that as the scale of the images increased so did the sense of either light or dark, so this change in scale is used throughout the pages of the book to get a sense of moving between the main streets into the small closes (see Figures 7.3 and 7.4). Occasionally the printed pages are also reversed and then an unprinted page included, implying the change from dark to bright

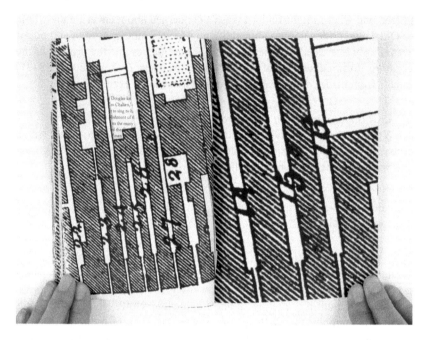

Figure 7.3 The cross hatching of the original map was used to gain a sense of the changing light levels

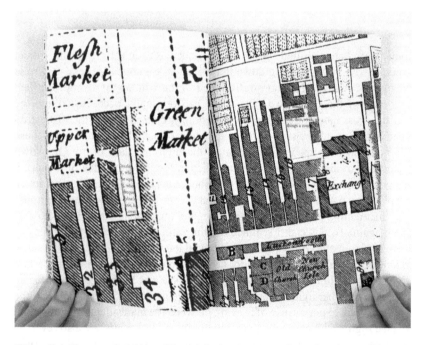

Figure 7.4 The cross hatching of the original map was used to gain a sense of the changing light levels

Figure 7.5 Blank pages were inserted to evoke the brightness on entering an open
courtyard

sunlight as one moves towards and into the open courtyards (see Figure 7.5).
As I was experimenting with this idea of negative and positive space I began to
cut away small, self-contained sections within the pages I was printing. In doing
so I was endeavouring to create a kind of fissure or break within the materiality of
place that would offer a kind of window into the past, enabling a shift in temporality and the viewing of a spectral presence.

   The idea for the Old Town project developed from an immediate sense of place
and a very basic physical interaction with the streets. Similarly, the design itself
developed through a physical interaction with the materials used and would not
have developed in this way without a 'hands on' engagement with paper and scal-
pel. Such immediate, often low-tech, types of experiments and prototyping bring
into play Schön's (1987) notion of reflection-in-action – a more immediate type
of analysis executed during the process of making. This form of reflection may
only be recorded in brief note form on a prototype, or perhaps not even that as one
may realise a potentially useful development during the actual process of making, enabling a swift change of direction that is recorded through the adjustment
of the design of the prototype itself rather than in words. Prototyping therefore
enables one to put theories relating to concepts and design into practice. Like
writing, which 'deepens our analytical endeavour' (Coffey & Atkinson 1996:
109), prototyping works similarly, by creating a physical, permanent form that
enables reflection and revision. Here, the material form of the work re-sites one

in place and allows further reflection on one's experience of place, thus remaking place on a second level (Pink 2015: 143). Consequently, through engaging in this material form of place-making on a second level, I began to get a clear idea of how I might construct something that would enable a reader to then remake place themselves on a third level through their interaction with the book (Pink 2015: 125). I realised that by once again using a French fold to create a 'hidden' space I could place texts within this space that relate to some of the events, buildings and characters that once brought to life these overcrowded streets, and that fragments of these texts would be visible through the cut away areas (see Figure 7.6). Even amongst the experimental and experiential research there is, and I would argue needs to be, room for the traditional. To produce work that is inspired by, and solely resides in, the experiential domain simply privileges one type of work over another, and misses out on a rich body of material that can both inspire and develop one's understanding in relation to place. Therefore, the *Old Town* book not only offers an embodied exploration of place, but also reveals texts that ground this experience in a historical context and reveals a sense of the spectral presence of absence.

In order to allow the reader to reveal the full text the edge of the French folded page is perforated. The perforations act in a similar way to Springgay, Irwin and Wilson Kind's (2005: 904) use of the /, as on tearing it and revealing the hidden

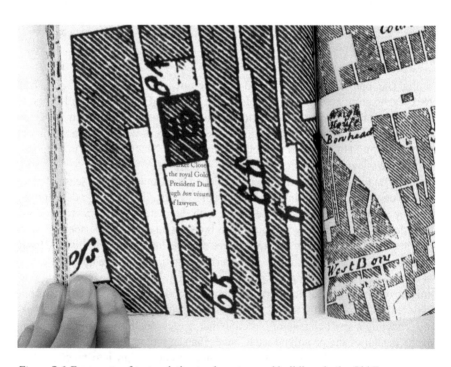

Figure 7.6 Fragments of texts relating to characters and buildings in the Old Town are visible through die cut sections

Figure 7.7 French folds are perforated so as to reveal internal texts

Figure 7.8 Separating the perforated pages doubles the pages within the book

texts, one immediately halves the single French folded page, but at the same time doubles it as it becomes two traditional pages. If the reader continues to separate all the pages the size of the book, and the number of pages, effectively double (see Figures 7.7 and 7.8). This alludes to the fact that place always has more to reveal, that stories of place are inevitably multiplicitous. In the same way that the Old Town of Edinburgh inspired me to explore in a physical, experiential way, through its design interventions, the book implicates the reader in a physical act of exploration in order to reveal its full contents. In our daily lives, we often rush through spaces, barely stopping to look at our surroundings – the desensualised spaces of the city often contributing to our smooth, uninterrupted path (Edensor 2008: 130). In terms of the reading experience, this is akin to flicking through the pages of a magazine, briefly gazing at fragments of colours and images as they speed by, but not stopping to read. However, in Edinburgh, the material qualities of the Old Town provoke a slower, multi-sensory experience. Similarly, the use of the French folds, the hidden texts, the perforations and the doubling of the space all slow the reader down, hinting at further things to be found, revealing fragments that encourage them to look further for what they may have missed.

This low-tech approach does not have to centre on pre-production prototypes. It can be seen as an accessible way of approaching the geo/graphic design process – one that is inclusive to both disciplines, rather than just graphic designers who may have access to specialist printing facilities. *Old Town* uses 80gsm ivory coloured paper purchased from a high street stationers and is printed using a £60 black and white laser printer. The colour of the paper echoes much of the brick I encountered in Edinburgh and the thinness of the paper leads to some show through, which adds to the sense that there is more to reveal behind each page. Binding has been done by hand, after consulting instructions from the internet. There is nothing here that is prohibitively expensive or difficult to get hold of. Perhaps the perception of a phrase like 'artists' book' conjures up the kind of limited editions that cost thousands of pounds and are held in the special collections of national galleries. However, this idea of a limited edition can be used in a positive way – equipment such as photocopiers, scanners, A4 paper and printers, found in most academic departments regardless of discipline, can be used to generate work that goes beyond the traditional use of such tools. Thus the physical development and construction of the book reveals much about the potential of the geo/graphic design process, whilst the book itself shows that, much like those discussed in the previous two chapters, the page has not been silenced by the advent of the mechanical printing process (Ingold 2007: 24). However, as the book foregrounds an embodied experience and the affective qualities of that, there is perhaps less in the way of specific insights about place available to the reader, rather it haunts the reader with 'vague imitations of the past' (Edensor 2005a: 829) and offers a disjointed, fragmented narrative. Yet, the process of walking and experiencing place prior to making the book did provide these types of insights.

The 'experience' of Edinburgh

Whilst engaging with the *Old Town* book is akin to exploring place alone, in reality the experience of exploring Edinburgh is more often than not shared with hundreds and thousands of others all treading the same, or very similar, well-worn and expertly narrated path. History – with a capital H – seems overtly present in Edinburgh. This is clearly a paradoxical statement as history is technically the past, yet Edinburgh's history seems to be formally memorialised at every opportunity. Plaques, statues, inlayed cobblestones – everywhere you turn another event or person is inscribed into or onto the landscape in some way. However, what do we mean by 'history'? Much of the history that is largely the focus of the various tours and experiences available for the tourist is that relating to the Old Town. If place is space that has been 'time thickened' (Crang 1998: 103) and palimpsestual in nature, then the heritage industry in Edinburgh seems to want to strip away the top 500 years of these layers to reveal what lies beneath. It seems an attempt to 'freeze' Edinburgh in time, to offer a totalising narrative of a time and place at the expense of what may have come before or afterwards. Perhaps Edinburgh's history seems all encompassing because it suffered no bomb damage in the Second World War. Much like Paris, this particular version of Edinburgh remains 'intact' with a huge number of historic buildings still standing and this seemingly untouched cityscape forms a large part of what visitors to Edinburgh wish to engage with.

As we have discussed previously, tourism in Edinburgh is a key part of the economy and the presence of a plethora of walking tours constantly moving throughout the city gives a real-time gauge as to how business is doing. Weaving in and out of the groups of people as you traverse the same streets in different directions you overhear snatches of other guides' stories as you pass. It's a complex, choreographed 'place-ballet' (Seamon 1980) that somehow manages to ensure no-one ends up in the same spot at the same time. The groups of people snake along, following their leader who usually carries some kind of identifiable flag or umbrella that is raised high above their head. This ensures you don't accidentally drift into a different group as you linger to take photographs or get distracted by something else in the vicinity of the stop you have just made for your guide to impart the next stage of their story. These tours run throughout the day, many of them are free, with the guides/storytellers working on a tips basis. The tours are offered in many languages and cover a range of different types – for example, general Old Town tours, ghost tours, 18+ adult themed walking tours, literary pub crawls, and Harry Potter tours. Some guides are older and Edinburgh born and bred. They seem keen to impart information in the way a school teacher might – slightly formally, with an emphasis on historical, social and political context. There are jokes and humorous asides, but the focus is on the facts and the stories, not the storytelling. Other, often younger, guides are doing this as a way to facilitate their world travels, having racked up stints in large cities across Europe. Not all from Edinburgh, these guides' emphasis is on the storytelling and their performative approach is more obvious. They get into character, they engage their whole body in the telling of the story – drawing in grimaces, gestures, and larger movements and engaging the audience as participants in their tales. Yet the fact

Figure 7.9 It is traditional for tourists to rub the nose of the bronze statue of Greyfriars Bobby

they aren't local, and that their focus is on the storytelling doesn't really change anything other than the delivery. Much like Greyfriars Bobby's shiny bronze nose (see Figure 7.9) or David Hume's shiny bronze big toe, the walking tour route through the Old Town is a well-worn one and the stories marking the stops along the way are simply learned verbatim like homework, then gradually personalised with the guide's particular style of delivery.

The stories told as part of these tours are seemingly not sanitised; we are told that the contemporary slang for being drunk – shit-faced – emanates from the practice of tenement residents emptying their chamber pots out of the windows above the street at ten o'clock each evening. This coincided with closing time for the pubs in the city, which saw the late night, often inebriated, drinkers attempting to find their way home through the warren of wynds and closes. When residents would shout 'gardyloo' – a corruption of the French 'gardez l'eau' which means 'mind the water' – those below would look up and the contents of the chamber pots would come to rest on their upturned faces. We are also treated to descriptions of the unsanitary living conditions, stories of body snatchers and cannibals, and the torture practices suffered by those convicted of a range of crimes, all of which are delivered with a sense of ghoulish glee. It is as if the people who suffered these conditions are so long dead we no longer worry about them – it is therefore not a tragic human story of squalor and deprivation, but entertainment.

There are also other 'historicising' tactics evident on the streets of Edinburgh. For example, typefaces are used on many of the shop front fascias that evoke the Gaelic script widely used from the sixteenth to eighteenth centuries, and traditional dress and historical costumes are evident on some tour guides and the bagpipe players who busk on the streets. Plaques sit within the entrances of many of the closes and wynds listing residents from previous times, and the boundary of the long since destroyed Tolbooth jail is marked out by a series of brass inlaid cobblestones (see Figure 7.10). Many pubs are named after some of the city's more infamous residents from times gone by – Deacon Brodie's Tavern, named after William Brodie, a deacon and cabinet maker who took to house burgling by night in order to fund his fondness for gambling and keep his two mistresses and several children; The Burke and Hare, named after William Burke and William Hare who committed a series of murders between 1827 and 1828 in order to take advantage of the lucrative business of selling corpses to Dr Robert Knox from the Royal College of Surgeons; and Maggie Dickson's named after Margaret Dickson or 'Half-hangit Maggie' who survived hanging in 1721 (see Figure 7.11). Thus spectral presences are everywhere, from the names of the infamous deceased above and others who are brought back to life on the plentiful tours. Several of these stories are repeated on the different tours I take, and therefore are likely to be repeated hundreds of times a month, or perhaps a week during the summer months. Repetition is 'crucial to the reproduction and evocation of memory' and such repeatedly told stories usher in 'the haunting apparition of the spectre' (Hill 2013: 391). Yet these repeated stories of Edinburgh's murky past and these overtly visible, often commercial, memorialised traces don't seem to evoke a sense of the past for me. In this well-trodden path with its overtly

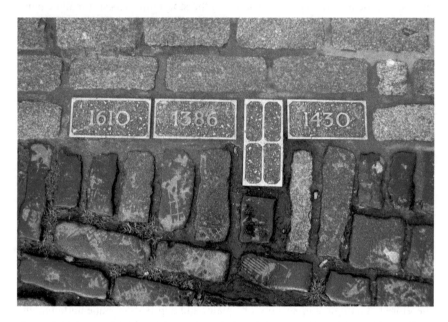

Figure 7.10 Bronze inlaid cobblestones mark the original boundary of the Tollbooth Jail

spectacularised and commercially driven narrative, any sense of an affective engagement with place is dulled, even though my body is still experiencing the same material realm of the Old Town. The stories remain literal for me – I hear the words, but they fail to conjure up a spectral presence. It is as if in being asked to engage my ears and mind, the rest of my body fails to follow and the multi-sensory realm of the Old Town becomes masked somehow. However, there is no doubt that even with the overt packaging of history for consumption by the tour-ists, the Old Town and its closes, wynds and courtyards are evocative. Its history seems very present; its absences are affective. Away from the tour groups and the planned routes and narratives I find myself running my hands over stone walls that are covered with the traces of time passing – worn grooves caused by I don't know what; the remaining holes from long since removed fixings for a fitting that is no longer there; and layers of flaking paint and plaster that reveals signs of, and signage from, a previous era (see Figure 7.12). I feel the cobble stones under my feet, my eyes blink as they react to the change from dark to light as I emerge from a dark close into a bright courtyard and my body senses the chilly dampness, and the closely built walls and low ceilings as I take yet another turn down a small passageway.

My engagement with the city, its hauntings and absent presences is therefore triggered primarily by a multi-sensory, physical experience rather than an aural and cerebral one. In the previous chapter, the affective experience of handling the materials within the pages of the book then triggered memories of my own

Figure 7.11 Maggie Dickson's pub, named after Margaret Dickson or 'Half-hangit
Maggie' who survived hanging in 1721

that resonated with these pieces of ephemera. The affective moment was therefore swiftly replaced by one that 'translated' this embodied, sensuous experience

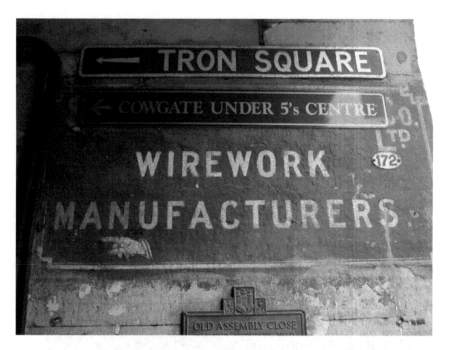

Figure 7.12 Layers of signage from different eras

into something more logical and literal, into something that made sense to me in the context of my own memories – the fragments took on a 'greater meaning' (DeLyser 2001: 27). Yet in Edinburgh, this embodied experience wasn't translatable – I didn't live in the 1700s, there was no logical extrapolation to make. Thus it remained a sensory experience – a kind of shiver with the awareness of absence, the sense of those who had trodden these streets before me.

Summary

Cities such as Edinburgh are continually emergent patchwork constructions that evidence multiple temporalities. They resist a totalising, ordering narrative of linearity, regardless of the construction of formally designated buildings that act as memorials or heritage sites in an attempt to fix the meanings of the past. Cities inevitably carry with them, and continue to produce, material traces that evoke past histories and spectral presences. As such, spectral presences resist this attempt to pin history down and the past continually emerges from within the present in ways that we often find evocative and affective. This type of haunting foregrounds absences that are produced between materiality and immateriality, and absence and presence, and this relational phenomenon is brought to life via our experiences within place. What prompts such moments is an embodied engagement with landscape that has resisted desensualisation. Much of Edinburgh's

Old Town retains elements of its seventeenth century construction. Thus we are literally in touch with place – the uneven surfaces of the cobblestoned roads and paths, the roughly hewn walls, and the dark, damp closes.

The *Old Town* book, and the process of making it, foreground this embodied, material experience and offers the reader an interactive space of exploration. The book includes texts relating to events that took place in the Old Town in previous centuries and the research and analysis offer insights as to how Edinburgh weaves the more commodified, formally constructed narratives of its past to provide the visitor with a particular view of place. The holistic approach of the Old Town book in the context of a geo/graphic approach therefore engages both mind and body directly, offering something to the debates about non-representational theory (Thrift 1996), and addressing the perceived divide between theory and practice (Nash 2000).

Bibliography

Barnes, A. (2012) 'Thinking geo/graphically: The interdisciplinary space between graphic design and cultural geography', *Polymath: An. Interdisciplinary Arts & Sciences Journal.* 2(3), pp. 69–84.

Campbell, T. (2016) 'Story of cities #10: how the dirty Old Town became enlightened Edinburgh', *The Guardian*, 29 March. Available at: www.theguardian.com/cities/2016/mar/29/story-of-cities-10-edinburgh-new-town-old-town-scottish-enlightenment-james-craig (Accessed: 7 December 2017).

Coffey, A. & Atkinson, P. (1996) *Making Sense of Qualitative Data: Complimentary Research Strategies*. London: Sage.

Crang, M. (1998) *Cultural Geography*. Abingdon: Routledge.

De Certeau, M. (1984) *The Practice of Everyday Life*. Berkeley: University of California Press.

DeLyser, D. (2001) 'When less is more: Absence and landscape in a California ghost town', in Adams, P. C., Hoelscher, S. & Till, K. E. (eds) *Textures of Place: Exploring Humanist Geographies*. Minneapolis: University of Minnesota Press, pp. 24–40.

DeSilvey, C. & Edensor, T. (2012) 'Reckoning with ruins', *Progress in Human Geography*. 37(4), pp. 465–485.

Dirksmeier, P. & Helbrecht, I. (2008) 'Time, non-representational theory and the "performative turn"—Towards a new methodology in qualitative social research', *Forum: Qualitative Social Research*, 9(2), Art. 55. Available at: www.qualitative-research.net/index.php/fqs/article/view/385/839 (Accessed: 7 December 2017).

Dittmer, J. (2014) 'Geopolitical assemblages and complexity', *Progress in Human Geography*. 38(3), pp. 385–401.

Edensor, T. (2012) 'Vital urban materiality and its multiple absences: The building stone of central Manchester', *cultural geographies*. 20(4), pp. 447–465.

Edensor, T. (2008) 'Walking through ruins' in Vergunst, J. L. & Ingold, T. (eds) *Ways of Walking: Ethnography and Practice on Foot*. Abingdon: Routledge, pp. 123–141.

Edensor, T. (2005a) 'The ghosts of industrial ruins: Ordering and disordering memory in excessive space, *Environment and Planning D: Society and Space*. 23, pp. 829–849.

Edensor, T. (2005b) 'Waste matter – the debris of industrial ruins and the disordering of the material world', *Journal of Material Culture*. 10(3), pp. 311–332.

166 *An embodied, affective experience of place*

Edinburgh Tourism Action Group. (2016) *Tourism in Edinburgh: Key Figures*. Available at: https://www.etag.org.uk/wp-content/uploads/2016/11/Facts-and-Figures-2016-Final. pdf (Accessed: 7 December 2017).

Garrett, B. (2011) 'Assaying history: creating temporal junctions through urban Exploration', *Environment and Planning D: Society and Space*. 29, pp. 1048–1067.

González, J. (1995) 'Autotopographies', in Brahm, G. & Driscoll, M. (eds) *Prosthetic Territories: Politics and Hypertechnologies*. Boulder: Westview Press, pp. 133–150.

Hill, L. (2013) 'Archaeologies and geographies of the post-industrial past: Landscape, memory and the spectral', *cultural geographies*. 20(3), pp. 379–396.

Ingold, T. (2007) *Lines: A Brief History*. Abingdon: Routledge.

Maclear, K. & Testa, B. (eds) *Life Style: Bruce Mau*. London: Phaidon.

Maddern, J. F. & Adey, P. (2008) 'Editorial: Spectro-geographies', *cultural geographies*. 15(3), p. 291–295.

Massey, D. (2005) *For Space*. London: Sage.

Massey, D. (1994) *Space, Place and Gender*. Minneapolis: University of Minnesota Press.

McMillan, A. & Hyslop, E. (2008) 'The City of Edinburgh – landscape and stone', *ICOMOS 6th General Assembly and International Scientific Symposium*, Quebec, Canada, 29 September – 4 October. Available at: https://www.icomos.org/quebec2008/ cd/toindex/77_pdf/77-KoCo-73.pdf#search=%27%27 (Accessed: 7 December 2017).

Meier, L. (2013) 'Encounters with haunted industrial workplaces and emotions of loss: Class-related senses of place within the memories of metalworkers', *cultural geographies*. 20(4), pp. 467–483.

Meier, L., Frers, L. & Sigvardsotter, E. (2013) 'The importance of absence in the present: Practices of remembrance and the contestation of absences', *cultural geographies*. 20(4) pp. 423–430.

Meyer, M. (2012) 'Placing and tracing absence: A material culture of the immaterial', *Journal of Material Culture*. 17(1), pp. 103–110.

Moran, J. (2004) 'History, memory and the everyday', *Rethinking History*. 8(1), pp. 51–68.

Nash, C. (2000) 'Performativity in practice: Some recent work in cultural geography', *Progress in Human Geography*. 24(4), pp. 653–664.

Nora, P. (1989) 'Between Memory and History: Les Lieux de Mémoire', *Representations*. 26, pp. 7–24.

Old and New Towns of Edinburgh (no date) Available at: http://whc.unesco.org/en/list/728/ (Accessed: 7 December 2017).

Paterson, M. (2011) 'More-than visual approaches to architecture: Vision, touch, technique', *Social & Cultural Geography*. 12(3), pp. 263–281.

Pink, S. (2015). *Doing Sensory Ethnography*. 2nd edn. London: Sage.

Quiring, B. (2010) 'A fiction that we must inhabit: Sense production in urban spaces according to Alan Moore and Eddie Campbell's *From Hell*', in Ahrens, J. and Meteling, A. (eds) *Comics and the City: Urban Space in Print, Picture, and Sequence*. London: Continuum, pp. 199–213.

Schön, D. (1987) *Educating the Reflective Practitioner*. San Francisco: Jossey-Bass.

Seamon, D. (1980) 'Body subject, time-space routines, and place ballets', in Buttimer, A. & Seamon, D. (eds) *The Human Experience of Space and Place*. London: Croom Helm, pp. 148–165.

Springgay, S., Irwin, R. I. & Wilson Kind, S. (2005) 'A/r/tography as living inquiry through art and text', *Qualitative Inquiry*. 11(6), pp. 897–912.

Thrift, N. (1996) *Spatial Formations*. London: Sage.

Conclusions

Introduction

Rather than use this conclusion to simply sum up and reiterate the key points from previous chapters, I want to use it to look at three aspects more broadly; the need for interdisciplinarity and collaboration; the potential of design and the geo/graphic approach; and how we might look to the future in relation to the use of creative methods in the context of the understanding and representation of everyday life and place. As we have seen, geographers are collaborating with artists in particular, and much interesting work is being done in relation to the use of creative methods, and indeed, collaborative and interdisciplinary approaches are becoming the norm in many contexts. Currently, this increase in, and emphasis on, interdisciplinarity is being driven by the realisation that many of the problems we face in the world are complex and are more often than not affected by multiple causes, and thus need to be understood from multiple perspectives before a solution might be developed.

That geographers have been drawn to art practice and artists is perhaps unsurprising as one of the roles contemporary art plays within society is to challenge our thinking, to engage with social, cultural, economic, environmental and political issues and provoke action and awareness – that the artist often acts as an activist or agitator is well known. Yet, less is known of design in this context and critical design, for example, has only recently been positioned as a field in its own right (Malpas 2017). However, design is inherently collaborative and interdisciplinary (McDermott, Boradkar & Zunjarwad 2014: 1), and could be said to occupy a space between the humanities and sciences (Rochfort 2002: 159) so clearly it does have the potential to contribute to such critical collaborative endeavours.

Whilst an engagement with design is one area that we might see as a way to further develop such interdisciplinary collaborations involving the use of creative methods, a further one involves looking at the context in which the creative methods are being used. Currently, much work – and I include this volume in that – looks at the field from a methodological perspective, focusing less on the context in which such projects are developed. However, for some, those engaging with creative methods need to engage with a critical dimension much more clearly (de Leeuw & Hawkins 2017). The opportunity to 'explore the politicalities of geographers' creative expressions' can clearly be found within geography's 'disciplinary history and existing critical and theoretical outlooks' (de Leeuw & Marston 2013: xxi), yet for some reason this doesn't seem to be occurring as it could. However, criticality can also be found in collaboration with art and design, and it is to this type of interdisciplinary approach that we first turn.

The need for interdisciplinarity and collaboration

Currently, there is an increasingly widespread interest in interdisciplinarity in both higher education and the workplace, and there exists an underpinning belief that it is foundational to both education and research (Repko, Szostak & Phillips Buchberger 2017: 4). The driving force for this emphasis on interdisciplinarity is the recognition that the majority of the challenges we face – at both local and global levels – are complex and therefore cannot be solved by those working within a single discipline. Such problems have been described by Rittel & Webber (1973) as 'wicked', in that they are extremely difficult to both map out and solve. They are underpinned by multiple causes; can often be a symptom of other issues; have no obvious solution or logical conclusion; and therefore resist resolution by traditional processes. Given that 'problem understanding and problem resolution are concomitant to each other' (Rittel & Webber 1973: 161), it is therefore clear that a single disciplinary, silo-like approach is unlikely to be suitable. One of the key benefits of interdisciplinarity is the propensity for new knowledge to be discovered at the 'borderlands between established disciplines and fields' (Repko, Szostak & Phillips Buchberger 2017: 11). Indeed, de Leeuw & Hawkins' findings within creative geographic collaborations corroborate this: 'The boundary-crossing works and collaborations we have participated in and observed offer important opportunities to create new spaces and modes for thinking about and expressing space, place, and human relationships with and within the world' (de Leeuw & Hawkins 2017: 305). There is no doubt that geographers collaborating with artists, or those geographers utilising creative methods in ways that are unfamiliar within the discipline, have had productive experiences, enabling many 'to ask awkward questions' of their 'own conventions and accepted working practice' (Foster & Lorimer 2007: 427–428). It is perhaps unsurprising that a focus on art has been established within such interdisciplinary collaborations, as the discipline of geography is able to build on previous analytical engagements with the visual in the context of art. Indeed, art plays an important role in in questioning our place in the world and challenging many of the social, cultural, environmental, economic and political scenarios that unfold within it. Even though, as we will discuss below, there may be questions as to why such critical approaches are not undertaken more often within such creative work by geographers, it is therefore perhaps not surprising that collaboration with art practice and artists themselves might be seen as a route to engaging with such issues.

However, design too has developed a critical and speculative approach. Originally identified and established within industrial and product design, such an approach seeks to reposition the role of design and the designer away from one that could be described as service-led, to one where design is used 'to mobilise debate and inquire into matters of concern' (Malpas 2017: 1–2). Thus, such 'critical design practice is positioned as a form of socially and politically engaged activity and creative activism' (Malpas 2017: 6). Design, and more importantly designers, also offer a further added benefit, in that much of their practice, particularly within the workplace, is inherently collaborative and interdisciplinary (McDermott, Boradkar & Zunjarwad 2014: 1). As discussed in Chapter 4, a design thinking approach is now often used in many contexts beyond design – in business and in health, for example. It offers an opportunity to address complex or 'wicked' problems in a way that 'can help generate creative solutions' whilst working within 'cross-functional teams' (McDermott, Boradkar & Zunjarwad 2014: 1). So, whilst a focus on interdisciplinarity itself might not be seen as a profoundly new or interesting insight, it is the connection to design that is particularly important in this context.

The potential of design and a geo/graphic approach

As we have seen above, within other disciplines and outside of the academy, design is recognised as having a vital role to play within many of the complex challenges facing the world today and design practice is by nature collaborative. Design's roots within the service industry and its transparent process position it as distinct from 'self-consciously lived, artistic expression' (Tolia Kelly 2012: 139). Tolia-Kelly is one geographer who does posit an argument for shifting the focus away from art and framing the use of creative methods within visual culture – going on to cite design as an area that might be productive for the positioning of a 'visual research edge' (2012: 139). Within the UK, the Arts and Humanities Research Council (AHRC) has designated design a strategic priority area and recently appointed a Design Leadership Fellow. The Design Leadership Fellow will assist in shaping new initiatives that centre on design and, in particular, 'how research and an increased understanding of the design process can showcase the profound difference that Design can make to societies' (*New Fellows help provide leadership in Arts and Humanities Research Council priority areas* 2016). Design with a 'capital D' obviously encompasses a multitude of sub-disciplines, including areas such as service design, product design, and even engineering, yet I would argue that across this diverse territory there is an agility to design thinking that is clearly well-placed to contribute to geography's creative turn. For example, design has been described as occupying a 'third area between the humanities and science' and it has been noted that a 'designerly way of knowing' combines skills from both (Rochfort 2002: 159). Yet often, those studying design, or perhaps even practicing it, are unable to locate or recognize themselves within this field (Swanson 1994: 59). In fairness, the discipline of design is still extremely youthful in terms of the academic context and design research is still an emerging field. So from this perspective, a geo/graphic approach will also prove useful to designers as it contextualises the creative approach within fields that are likely to be unfamiliar to them. As Jellis (2015: 369) and Tolia-Kelly (2012: 137) both point out, little has been written about the benefits to artists working collaboratively with geographers, and whilst this volume doesn't address that, what it does do is offer artists and designers an opportunity to engage more deeply with theory and practice from geography and anthropology. Critical design is also still at an early stage in its development, with Malpas (2017: 3) suggesting that up until now, the field has remained relatively insular, often only engaging with other artists and designers. In order for this to change, his view is that it should be challenged from beyond the discipline of design

> . . . ultimately, for critical design practice to be truly critical and offer value in extending the agency of design, challenge disciplinary thinking, and possibly even effect change, it must not be above critique itself.
>
> (Malpas 2017: 3)

Thus positioning oneself outside of and beyond one's own territory enables one to begin to frame practices that may have been instinctive, or learnt by trial and error, within the rich traditions of social science research and practice. It also enables a shift away from 'designer as author' towards 'designer as researcher' (see Barnes 2012) and repositions design from a service-led field to one that can take an equal part in the collaborative work that engages with complex problems. For graphic design in particular, this opens up the possibilities to reposition it as less of a communicative tool and more as 'an instrument for the production and communication of knowledge' (Mermoz 2006: 77).

In terms of the geo/graphic process specifically, first it offers geographers and graphic designers an opportunity to learn more about each other's disciplines, theories and practices, and in turn paves the way for future collaborations. In doing so, it dispels the notion that the page has been silenced by the mechanical process of print (Ingold 2007: 24) and that print based representations are therefore unable to contend with ongoing, relational aspects of place. Utilising a geo/graphic approach, the book is revealed as having the potential to create a re/presentation of place that is every bit as open, engaging and affective as versions of place developed using moving image or sonic methods. As we will discuss further below, the impact agenda within higher education research will perhaps offer a challenge to traditional forms of academic publishing, but currently, many departments outside of art and design are conservative in their approach to what has been submitted within research assessment exercises (Nash 2013: 54). Thus research outputs in traditional printed formats of books, book chapters and journals are likely to remain an established part of the academic culture for the foreseeable future. So developing approaches to traditional publishing that 'enliven conventional printed formats' (Nash 2013: 54) and create 'experiential texts' (Pink 2013: 86) offers an opportunity to reinvigorate the field and meet the challenge posed by the more-than-representational (Lorimer 2005) aspects of everyday life and place and of doing sensory ethnography (Pink 2015: 4). However, whilst calls to develop engaging texts (Richardson & St. Pierre 2005) or develop a 'geopoetics' (Springer 2017) are increasing, novel forms of output, such as those using a geo/graphic approach, should not be judged on novelty alone (Nash 2013: 54) – work utilising creative methods needs to be both experiential *and* scholarly (Pink 2015: 40).

Second, a geo/graphic approach offers an overarching methodology that in drawing from multiple disciplines develops an approach that is adaptable and transferable. It is a methodology that can be tailored to particular contexts, projects and collaborations, and therefore one that, whilst offering relatively broad parameters within which to work, enables a more specific set of methods – in terms of those used to understand *and* represent place – to be developed as part of the research process. This is evident in the context of the projects discussed in Chapters 5, 6 and 7, in which the scale of the sites are diverse, the type of approach ranges from the autoethnographic to the participatory, and the research contexts range from ideas of home, personal possessions and collections; food, multi-culturalism and belonging; to memories and the affective dimensions of material traces within the environment. However, whilst the focus of this book is primarily to add to the current body of work around the methodological approach to, and application of creative methods, there is naturally also a need to look to the future, to discuss how the use of, and the approach to, creative methods might develop further.

Looking to the future

There is no doubt that the interest in creative and visual methods is increasing – the plethora of conferences, journal articles and research centres focusing on aspects of this emerging field are testament to this. The various 'crises' and 'turns' encountered within geography and the social sciences over the past few decades have perhaps inevitably led us to a point where the 'more-than-representational' (Lorimer 2005) aspects of everyday life and place have begun to take centre stage. In these twenty-first century engagements with place 'knowing through practice' (Wenger 1998) has become key, with place experienced and understood through the body. A focus on multi-sensoriality, atmospheres and the affective dimensions of everyday life and place have offered a re/presentational challenge that has seen geographers explore technologies such as film and sound so as to avoid the assumed representational fixity of print. Some have suggested there is a danger with this, that this upsurge in experimental methods may draw researchers into methodological novelty for novelty's sake (Shaw, DeLyser & Crang 2015) and whilst this may be the case, it is only natural to explore the boundaries of such new territory. However, what might be problematic is the critique that such creative geographies are, as yet, largely non-politicised (de Leeuw & Hawkins 2017).

De Leeuw & Hawkins (2017: 308) are curious as to why within this creative turn: 'many geographers are producing creative work and undertaking creative practices with little or no explicit reflection on or explanation of the politics of their work or the works' political implications'. They are perhaps correct that this is a seemingly odd state of affairs given that the series of challenges to the status quo within geography began with the realisation that dominant narratives within the discipline were excluding marginalized voices and were contributing to a continued imbalance of power. Currently, much of the work engaging creative methods is undertaken 'in order to create geographic understandings about the world', 'or to reflect on geographical scholarship' (de Leeuw & Hawkins 2017: 307). De Leeuw & Hawkins make the observation that in the main, when geographers themselves are producing creative works as part of their research practice, critical intervention or reflection is limited. However, when geographers are collaborating with artists, activists and/or communities, and co-producing creative work, this often results in work that is both creative and critical (de Leeuw & Hawkins 2017: 307). One could argue that this is perhaps an unfair comparison given that the context of a project involving artists, activists and/or the community is likely to have a political or critical dimension. For example, an activist by very definition believes in social and/or political change and strives to achieve this; similarly, participatory work with many communities is a way of giving a voice to those who may have little say over decisions being made on their behalf; and art itself does not exist in a vacuum of aesthetics, but plays a key role in questioning social, cultural, economic and political concerns. Yet, the tradition of critical geography is underpinned by similar concerns, so why are so few geographers producing work through the use of creative methods that utilises its powerful political potential for challenging and changing 'not just how we conduct research and create knowledge, but also how we live in the world' (de Leeuw & Hawkins 2017: 307)?

I suggest there are possibly two things occurring here that contribute to this perceived lack of critical and politicised creative output. Firstly, undertaking what is essentially interdisciplinary work as an individual can be challenging. As Nash (2013: 53) notes, achieving 'acceptance as a creative practitioner' can be difficult as artists (and designers) are likely to have undertaken a degree in their subject and developed their approach and body of work over a period of time. To legitimise or validate ones work as 'art' may therefore seem a difficult proposition and may lead to the researcher focusing on the method and material itself as this is the aspect of the work they are least familiar with. In doing so, they perhaps overlook what is most familiar to them – in this case, their critical geographical approach that seeks to question and challenge aspects of the social, cultural economic and political landscape of everyday life and place. Interestingly both Jellis (2015: 369) and Tolia-Kelly (2012: 137) note that little has been written from the opposite perspective – that of the artist (or designer) attempting to inhabit a geographical world, and that similar issues around a lack of confidence in claiming such territory are likely. Whilst it is difficult to generalise, as each individual's experience is likely to be different, my own is that, at times, undertaking interdisciplinary work as an individual has the potential to leave one vulnerable to the 'imposter syndrome' and a sense that one is somehow fraudulently occupying territory that is beyond the bounds of one's own discipline. However, given that the research projects geographers might develop are increasingly likely to feature complex problems, I would also suggest that to engage in work of an interdisciplinary nature without collaborating with others from beyond their own disciplines is short sighted. It would seem inevitable that this is less likely to lead to an outcome that engages fully with the multitude of issues that come together to form such a 'wicked' problem and therefore less able to 'challenge and change the status quo' (de Leeuw & Hawkins 2017: 308).

Secondly, as with any emergent field, much work is done at the outset that is methodological in nature – as was the case with the projects that formed the basis of the previous three chapters. This does not mean that any work necessarily eschews a political context, more that the emphasis of the study is on the methods themselves. As both Pink (2015: xii–xiii) and Davies & Dwyer (2010: 95) have stated with regard to such creative methods, we are still at a stage when a wide range of disciplines could benefit from a clearer, more explicitly discussed focus on exactly how such creative methods are undertaken and there are many examples of this (for example Garrett 2011; Lorimer 2010; Gallagher & Prior 2014), to which we might add this volume. This methodological focus perhaps also offers an answer to those who fear that researchers may be utilising novel methods for novelties sake (Shaw, DeLyser & Crang 2015). At this stage, the parameters of the discipline within the context of creative methods are being tested purposefully, with researchers exploring and experimenting with unfamiliar practice-based methods in the context of theory.

One further context is also likely to play a part in the increase in creative methods and that is the increased focus on impact in the assessment of academic research. Particularly in higher education in the UK, the impact of research is now 'measured' beyond journal article metrics and impact looks at how a piece of research might engage the public or be perceived as useful. For some, this has led to a concern that the positioning of the visual may simply be as 'an accessible mode of research' (Tolia-Kelly 2012: 135), and that creative methods will be used uncritically, for their own sake (de Leeuw & Hawkins 2017: 319), as part of a 'new orthodoxy' (Tolia-Kelly 2012: 137). This challenge to the 'primacy of books and papers in academic knowledge production' (Nash 2013: 54) is therefore likely to see an increase in methods that position research as more accessible and situated within realms that are beyond that of the traditionally conservative academic publishing world – in the gallery or community, for example. However, to create lasting impact or change, creative methods need to be harnessed in the context of critically engaged projects that are interdisciplinary in nature, and therefore able to address the complex scenarios that underpin everyday life and place.

Whilst impact and public engagement are increasingly important, academic publishing is unlikely to reduce in importance and the likelihood is that academics will develop research projects that can be disseminated in a range of ways. Currently, de Leeuw & Hawkins (2017: 309) state that many of the journals they describe as 'expressly critical', such as '*ACME, Antipode, Human Geography* or *Gender, Place and Culture*' currently rarely publish work undertaken by geographers that utilises a creative form of re/presentation. Given that the use of creative methods, although increasing, cannot yet be said to be spread across the global discipline consistently; that there still remain questions as to how to assess such work (Hawkins 2012: 65); and that the explicit remit of many of these journals makes the inclusion of a creative piece without accompanying analytical text difficult (see Butz 2011: 278), this is not surprising. Perhaps, therefore, we might see this limited critical engagement with creative methods as a phase within the development of this new approach. Once parameters are more clearly defined; once the methodological literature articulates more clearly the nuances of various approaches; and once these approaches become more established throughout the discipline as a respected and unquestioned form of qualitative research, the critically questioning, political, creative work is likely to follow.

To conclude, given the increased need for interdisciplinary work and the focus on impact within the academic research landscape, my sense is that creative methods still have much to contribute within geography and the social sciences. Whilst art practice may continue to be at the forefront of these developments, that design will begin to make a contribution within the field is perhaps inevitable. There is an increasing awareness and uptake of design thinking methods beyond universities, design schools and design agencies; a designerly approach is inherently collaborative and interdisciplinary; design and design research continues to develop within and outside of the academy; and movements such as critical design are becoming more established and are beginning to be disseminated more widely. Whilst this volume has focused on graphic design and visual communication specifically, design has a multitude of sub-disciplines, all of which have the capacity to utilise the design process and creative thinking within the context of complex interdisciplinary projects. Such projects will not only increase our understanding of everyday life and place, but will also begin to articulate critical responses that also seek to effect change, for central to design and designers is the remit and capacity to problem solve.

Bibliography

Barnes, A. (2012) 'Repositioning the graphic designer as researcher', *Iridescent*. 2(1), pp. 3–17.

Butz, D. (2011) 'The bus hub: Editor's preface', *ACME: An International Journal for Critical Geographies*. 10(2), pp. 278–279.

Davies, G. & Dwyer, C. (2010) 'Qualitative methods III: animating archives, artful interventions and online environments', *Progress in Human Geography*. 34(1), pp. 88–97.

de Leeuw, S. & Hawkins, H. (2017) 'Critical geographies and geography's creative re/turn: Poetics and practices for new disciplinary spaces', *Gender, Place & Culture*. 24(3), pp. 303–324.

Foster, K. & Lorimer, H. (2007) 'Cultural geographies in practice: Some reflections on art-geography as collaboration', *cultural geographies*. 14(3), pp. 425–432.

Gallagher, M. and Prior, J. (2014) 'Sonic geographies: Exploring phonographic methods', *Progress in Human Geography*. 38(2), pp. 267–284.

Garrett, B. (2011) 'Videographic geographies: Using digital video for geographic research', *Progress in Human Geography*. 35(4), pp. 521–541.

Hawkins, H. (2012) 'Geography and art. An expanding field: Site, the body and practice', *Progress in Human Geography*. 37(1), pp. 52–71.

Ingold, T. (2007) *Lines: A Brief History*. Abingdon: Routledge.

Jellis, T. (2015) 'Spatial experiments: art, geography, pedagogy', *cultural geographies*. 22(2), pp. 369–374.

Lorimer, H. (2005) 'Cultural geography: The busyness of being "more than representational"', *Progress in Human Geography*. 29(1), pp. 83–94.

Lorimer, J. (2010) 'Moving image methodologies for more-than-human geographies, *cultural geographies*. 17(2), pp. 237–258.

Malpas, M. (2017) *Critical Design in Context: History, Theory and Practice*. London: Bloomsbury.

Marston, S. A. & de Leeuw, S. (2013) 'Creativity and geography: Toward a politicised intervention', *The Geographical Review*. 103(2), pp. iii–xxvi.

McDermott, L., Boradkar, P. & Zunjarwad, R. (2014) 'Interdisciplinarity in design education: Benefits and challenges', In proceedings of *IDSA International Conference & Education Symposium*. Austin, Texas, USA, 13–16 August.

Mermoz, G. (2006) 'The designer as author: Reading the city of signs—Istanbul: Revealed or mystified?', *Design Issues*. 22(2), pp. 77–87.

Nash, C. (2013) 'Cultural geography in practice' in Johnson, N., Schein, R. H. & Winders, J. (eds) *The Wiley-Blackwell Companion to Cultural Geography*. Chichester: John Wiley & Sons, pp. 45–56.

New Fellows help provide leadership in Arts and Humanities Research Council priority areas (2016) Available at: www.ahrc.ac.uk/newsevents/news/new-priority-area-fellows/ (Accessed: 21 January 2018).

Pink, S. (2015) *Doing Sensory Ethnography*. 2nd edn. London: Sage.

Pink, S. (2013) *Doing Visual Ethnography*. 3rd edn. London: Sage.

Repko, A. F., Szostak, R. & Phillips Buchberger, M. (2017) *Introduction to Interdisciplinary Studies*. 2nd edn. Thousand Oaks: Sage.

Richardson, L. & Adams St. Pierre, E. (2005) 'Writing: A method of inquiry' in Denzin, N. & Lincoln, Y. (eds) *Handbook of Qualitative Research*. Thousand Oaks, California: Sage, pp. 959–978.

Rittel, H. W. J. & Webber, M. M. (1973) 'Dilemmas in a general theory of planning', *Policy Sciences*. 4(2), pp. 155–169.

Rochfort, D. (2002) 'Making connections: Design and the social sciences', in Frascara, J. (ed.) *Design and the Social Sciences: Making Connections*. London: Taylor Francis, pp. 157–165.

Shaw, W., DeLyser, D. & Crang, M. (2015) 'Limited by imagination alone: Research methods in cultural geographies', *cultural geographies*. 22(2), pp. 211–215.

Springer, S. (2017) 'Earth Writing', *GeoHumanities*. 3(1), pp. 1–19.

Swanson, G. (1994) 'Graphic design education as a liberal art: Design and knowledge in the university of the "real world"', *Design Issues*. 10(1), pp. 53–63.

Tolia-Kelly, D. (2012) 'The geographies of cultural geography II: Visual culture', *Progress in Human Geography*. 36(1), pp. 135–142.

Wenger, E. (1998) *Communities of Practice: Learning, Meaning and Identity*. Cambridge: Cambridge University Press.

Index

Milton Keynes UK
Ingram Content Group UK Ltd.
UKHW040055071024
449327UK00019B/577